Pests & Diseases

Essential know-how and expert advice for gardening success

Pests & Diseases

CONTENTS

Lily beetles are one of many common insect pests that can spoil the appearance of ornamental plants and damage edible crops if left unchecked.

PLANT PROBLEMS

The health of plants is important to every gardener, but our attitudes to pests and diseases have shifted in recent years. We have become more tolerant of minor problems and found ways to garden in closer harmony with nature, reducing our use of chemical pesticides, which can harm beneficial wildlife. Spotting problems early is crucial to effective control, so learning to recognize the tell-tale signs of pests, diseases, and disorders is always worthwhile.

ADOPT A POSITIVE APPROACH

We grow plants to enhance our surroundings or to produce generous harvests of crops. Pests and diseases can thwart these ambitions and it is tempting to turn to chemicals to remedy problems, but in recent decades it has become clear that pesticides have negative impacts on wildlife and the environment.

Fortunately, there are positive alternatives. By building your knowledge about pests and diseases you can identify the best remedies, and work with nature – by, for example, promoting soil health and encouraging beneficial wildlife into your garden – to grow healthy and beautiful plants.

Diverse plantings help to avoid problems with pests and diseases.

PERFECTLY IMPERFECT

Don't feel intimidated by the immaculate blooms and unblemished rows of crop plants that you see on the pages of gardening magazines. Gardening should be a relaxing pursuit. Learn to tolerate minor damage that most pests and diseases inflict on healthy plants and accept that nibbled leaves or a few marks on your homegrown produce are nothing to worry about. Opt for planting styles full of diversity, where a bit of damage is less noticeable and problems are likely to affect only a few of the plants. Try roses in a mixed border with bulbs and perennials rather than a traditional rose bed, or a potager-style planting where edible crops are mixed with ornamentals.

ENCOURAGE A NATURAL BALANCE

Successful gardening is about creating an environment where a variety of organisms can flourish alongside your plants, rather than trying to wipe out the "bad guys" branded as pests and diseases. Promoting this diversity, by limiting or avoiding the use of chemical controls and providing sources of food and shelter, repays you by bringing in beneficial creatures, which naturally limit pest populations on your plants. Adding plenty of organic matter to your soil each year will also attract soil-dwelling creatures to break it down as they feed, boosting soil health and producing robust plants that are more resistant to disease. Every garden is potentially a resource to share with wildlife, filled with year-round blooms for pollinating insects, berries and bugs for birds, and shelter for all kinds of creatures. It makes good sense to welcome them in.

Bumblebees are valuable pollinators, and can easily be attracted into your garden by planting colourful flowers.

Tap into local gardening groups and online forums for targeted advice.

GET LOCAL KNOWLEDGE

Dealing with pests and diseases effectively depends on pinpointing the problem. This is no easy matter. Many varied organisms attack plants, often producing similar symptoms. This book will help identify the agents responsible, but there's no substitute for experience and local knowledge.

Start by finding out which pests and diseases trouble your neighbours, fellow allotment-holders, or even local farmers.

This will narrow down the most likely candidates and suggest preventative measures to stop the pests and diseases before they take hold. Use local information to guide your choices when planting, by selecting ornamental and edible plants less susceptible to problems in your area, or even timing the planting of vegetable crops to avoid them. Knowing what to watch out for will also allow you to spot symptoms early and deal with pests and diseases swiftly before any major damage is done.

LEARN FROM YOUR LOSSES

Even in the most successful gardens, some seedlings will be eaten by slugs, fruit crops will be damaged by birds, and a few plants will succumb to frost in a cold winter. Far from being a sign of failure, these losses are an inevitable part of gardening, providing a valuable way to learn how to better protect your plants from harm next year, or a useful indicator of plants that might be unsuitable for your soil conditions and climate. The key to taking a positive away from a disappointment is always to try to identify the cause of the problem, whether that's a pest, a disease, or a mistake when sowing or planting, to increase your horticultural knowledge and avoid it being repeated.

Using netting over blackcurrants keeps the soft fruits protected from hungry birds.

Tolerating a degree of damage is part of a sustainable pest-control approach.

WHAT ARE PLANT PESTS?

A pest is any animal – from a tiny insect to a large mammal or bird – that damages the growth of garden plants. Most pests live above ground, feeding on leaves, flowers, and fruits, or sucking sap, which means that they rarely go unnoticed for long. Some, however, are soil-dwellers and are much harder to spot because they feed on underground roots, bulbs, and tubers. They can cause serious damage before symptoms become obvious above ground. Learning to recognize the pests in your garden enables you to nip most problems in the bud, or to prevent them from establishing at all.

INSECT FOES AND ALLIES

While many garden pests are insects, it is also true that insect activity is essential to every healthy garden. Some insects are vital pollinators, others help in the essential work of decomposition, and others still prey on garden pests, making them your allies. It pays to be able to distinguish friend from foe (see pp.28–29).

Insects are a diverse group of animals. Adult insects all have six legs and a body made up of three parts, and may have one or two pairs of wings or be wingless.

The main pest groups are aphids, beetles, whiteflies, and thrips. They may suck sap or simply munch their way through leaves, flowers, or stems.

It's not just the adults you need to watch out for, either. Many insects emerge from their eggs as caterpillars, grubs, or maggots, which feed on leaves, fruit, or roots before pupating into their adult form. For this reason, experienced fruit and vegetable growers often cover crops with netting or use other barriers to prevent egg-laying butterflies, moths, and flies reaching their plants.

Most familiar invertebrates, such as centipedes, millipedes, and woodlice, are neutral or beneficial to gardeners.

OTHER INVERTEBRATE GARDEN PESTS

Apart from insects, the most significant invertebrate pests are slugs and snails, which feed voraciously at night and during wet weather. They target leaves, stems, flowers, and tubers, often leaving plants looking ragged, rows of seedlings destroyed, or potato harvests spoiled with only slimy trails revealing the culprits.

Other notable invertebrate pests include tiny mites, which suck the sap of plants, damaging foliage or causing deformities such as blisters and galls on shoots, and eelworms (nematodes) – microscopic threadlike organisms that live in the soil. Some eelworm species will penetrate plant cells to feed and destroy tissues from within; others are harmless or are used as biological controls (see pp.42–43) to counter plant pests.

The caterpillar of the large white butterfly is a common pest of brassicas.

Lily beetles are easy to spot, but you may not recognize their dark larvae.

Rabbits can cause significant damage, even when present in small numbers.

MAMMALS AND BIRDS

More sizeable, mobile, and intelligent than invertebrates, mammals and birds can cause huge problems for gardeners. Grazers, like deer and rabbits, find the fresh shoots of ornamental and veg gardens irresistible, while wood pigeons love pea shoots and brassicas. Mice and rats are opportunists that will feed on planted or stored vegetable seeds and bulbs, while squirrels will also dig up bulbs and strip bark from trees. The digging of domestic cats, dogs, and chickens will ruin seedlings and damage established plants and lawns. Barriers are usually the best way to keep mammals and birds away from plants.

NEED TO KNOW

Check your plants regularly for these symptoms, which may point to pest damage:
- Holes, nicks, or tears in foliage.
- Infested or deformed shoot tips.
- Discoloured patches on leaves.
- Excessive wilting (caused by root damage).
- Tunnels or maggots in fruit and vegetables.
- Seedlings destroyed or badly damaged.

WHEN TO WATCH FOR TROUBLE

Many pest problems are seasonal because the life cycles of both pests and plants are closely linked to the weather, day length, and the availability of food.

Most insect pests appear in spring, and may reproduce very quickly as the temperature rises, so their numbers can explode as spring turns into summer. The life cycles of pest predators are linked to those of their prey, and it takes time for predator populations to increase to the point where they can bring pest numbers under control. Using insecticides will kill the pests but also wipe out predators, making the use of chemical controls counterproductive in many situations (see p.31).

Pest numbers fall as temperatures drop and growth slows in autumn. This is the time to check sheltered spots to find and eliminate overwintering pests (see below). Mammals and birds are most likely to cause problems in winter, simply because gardens provide food sources during cold weather.

Aphids multiply very fast, but are soon controlled by their natural enemies.

OVERWINTERING PESTS

Invertebrate pests are generally dormant during winter. Look for them among fallen leaves, in piles of logs or stones, or in crevices on shrub and tree bark. Pick off and kill hibernating snails and clusters of woolly aphids, or apply a winter wash containing plant oils to remove aphid eggs hidden in the bark of trees. Hanging bird feeders in trees can also encourage birds to eat overwintering pest eggs.

Snails may hibernate in cracks and crevices.

WHAT ARE PLANT DISEASES?

Diseases occur when microorganisms infect plants and cause damaging symptoms, often visible as spots, marks, or rots on different parts of the plant. Fungi are the most common plant pathogens, followed by viruses, and – less frequently – bacteria. Some pathogens only infect a small group of closely related plants, while others are less fussy, but disease is always more likely in plants already stressed by environmental factors or pest damage. Serious problems can often be prevented simply by providing plants with suitable growing conditions and maintaining good garden hygiene (see *pp.32–33*).

Grey mould commonly affects soft fruits, such as raspberries, strawberries, and grapes.

Colour breaks, as in this tulip flower, are sometimes considered attractive.

FUNGAL DISEASES

When you think of fungi, mushrooms and toadstools might spring to mind. These structures are the large "fruiting bodies" of networks of underground fungal strands. However, the majority of fungal diseases, including powdery mildew, rusts, and grey mould, are caused by much smaller organisms. The presence of fine, fluffy strands of fungal cells or powder-like spores usually makes them fairly easy to spot and helps to distinguish fungal diseases from viral and bacterial infections.

Most fungal infections begin by affecting a particular part of the plant, producing distinctive spots, blotches, powdery coatings, or mould on leaves, flowers, and stems and causing roots to blacken and rot. Once established, the fungal infection can spread through plant tissues to cause whole stems, or even entire plants, to die. Wet and humid conditions favour fungal growth and the transmission of spores, which spread infection, so ensure your soil is well drained, avoid splashing water on to foliage, and maintain good airflow around plants to help keep them healthy.

VIRUSES

Microscopic viruses infect a wide range of garden plants. Their size makes them impossible for gardeners to identify directly, so they are usually named for the symptoms they cause and the type of plant affected. Viruses will be present throughout an infected plant, but may only cause symptoms in one area, such as the leaves or flowers. Typically, leaves display yellow markings with distinctive mosaic, streaked, or mottled patterns, and may be distorted, puckered, or rolled. Flowers can also be malformed or have white streaks, known as "colour breaks". Growth of infected plants is often poor or stunted, but viruses rarely kill plant hosts. Some mildly affected plants continue to grow quite well and may even enter into cultivation as new ornamental varieties.

BACTERIAL DISEASES

Only a few bacterial diseases trouble garden plants, but those that do include bacterial canker and fireblight, which may lead to rapid damage of infected tissue and sometimes significant dieback if not caught early. These diseases are not always easy to distinguish from fungal infections, but classic signs that bacteria are the culprits include yellow halos around leaf spots, darkened, sunken areas on stems, and oozing from sites of infection. Affected roots and underground bulbs, rhizomes, or tubers will become discoloured and quickly rot down into a slimy residue, which usually has an unpleasant smell. Bacteria can spread in water droplets splashed up from the soil and may enter plants through damage caused by pests, pruning, or bad weather.

Fireblight is a serious bacterial disease that can kill apple and pear trees.

Bacterial canker affects stone fruit trees; sour sap oozes from lesions on branches.

SOURCES OF INFECTION

To offer your plants the best protection against diseases, you need to understand how pathogens live and proliferate, and the mechanisms by which they infect plants. Healthy plants have resistance to infection, but even a small wound caused by a pest, wind damage, or leaf scorch creates an easy entry point for disease. Films of water on the surface of the plant provide ideal conditions for many airborne fungi to take hold and multiply, while water droplets splashing up from the ground can bring soil-borne pathogens with them. Some fungi and bacteria can persist in the soil for years waiting for a suitable host, while others lurk on weeds, on contaminated pots and tools, among fallen leaves, in garden compost, or even on new plants bought or received as gifts. Viruses tend to spread by the transfer of sap from one plant to another during pruning, propagation, or visits by sap-sucking insect pests.

Water the soil beneath the plant; don't spray water on the leaves.

NEED TO KNOW
Watch out for conditions that make plants more prone to infection.
- Poor airflow: plants are too close together, badly pruned, or in a poorly ventilated greenhouse.
- Wet or waterlogged soil: reduces the supply of oxygen to roots and creates conditions for fungal rots.
- High humidity on leaves or in the air: facilitates the spread of fungal infections.

Remove leaves that have fallen off a diseased plant to prevent spread.

WHAT ARE PLANT DISORDERS?

Plants displaying symptoms that are not caused by pests or diseases may be affected by disorders triggered by other factors in the environment. These range from poor growing conditions to exposure to harmful chemicals. Weak growth and discoloured foliage are common signs that a plant is under one of these stresses, but disorders can also lead to a lack of flowering, spoiled fruit and vegetable crops, and even death, so it is important to take them seriously. The good news is that, once recognized, many disorders can be rectified, especially if spotted early.

Adding slow-release fertilizer to the compost of container-grown plants ensures that they can access vital nutrients.

A layer of organic mulch enriches the soil beneath plants with nutrients and helps to improve soil structure.

Magnesium deficiency in this potato plant has caused a pronounced yellowing between the leaf veins.

NUTRIENT DEFICIENCIES

Plants take up the mineral nutrients that they need for healthy growth from the soil via their roots. High quantities of nitrogen, phosphorous, and potassium (referred to by their chemical symbols, N, P, and K) are required, along with smaller amounts of calcium, sulphur, and magnesium, and trace quantities of iron, boron, and molybdenum.

Nutrient deficiencies can occur where plants have exhausted the supply of a particular nutrient – a common problem when they are grown in containers – or the mineral has been washed out of a sandy soil by heavy rain. At other times, nutrients may be present in the soil but are unavailable to plants because the soil is too dry or of an unsuitable pH (see p.24). If a particular nutrient is in deficit, growth slows, leaves become discoloured and deformed, and fruits may develop telltale marks. These symptoms are undesirable in both ornamental and edible plants; what's more, specimens stressed by nutrient deficiency are more vulnerable to attack by pests and diseases. Luckily, deficiencies are usually easily avoided by regularly improving soil with mulches of organic matter, or feeding plants with fertilizers (see pp.26–27) where a more immediate boost of nutrients is needed.

Avoid waterlogged container plants by adding broken pots to aid drainage.

Drought followed by heavy watering can cause fruits like tomatoes to split.

COMMON SYMPTOMS

No plant – however hardy – is resistant to disorders. To minimize problems, choose plants carefully to suit the conditions in your garden and plant them where they receive optimum levels of moisture and light (see p.24).

A lack of light can cause sun-loving plants to become pale and drawn, while intense summer sunlight can scorch dry, brown patches on delicate leaves. Insufficient water at the roots stresses plants so that their foliage wilts, and they are left susceptible to disease. Foliage usually recovers after watering or rain, but an inconsistent water supply can result in nutrient deficiencies and split fruits, roots, and bark, which in turn creates entry points for disease.

Waterlogged soil commonly occurs in containers with inadequate drainage; yellowing leaves and poor growth are early signs of trouble. Wet conditions encourage fungal diseases, which rot the roots or crown, so that the whole plant wilts or dies back and is unlikely to survive in the long term.

ATYPICAL GROWTH

Healthy stems sometimes grow in an atypical form. Herbaceous and woody plants may form flattened shoots with odd arrangements of flowers. This phenomenon, known as fasciation, may be caused by a mutation or by damage to the developing shoot. It doesn't harm the plant and doesn't usually recur, so can be pruned out if required.

Variegated shrubs and trees may be affected by reversion, where loss of the mutation for the foliage colour results in vigorous green shoots dominating unless they are removed promptly.

This variegated *Euonymus* plant has produced a whole reverted shoot.

PHYTOTOXICITY

Damage to plants caused by any chemical substance is called phytotoxicity. It can on occasion arise from unexpected sources that are difficult to identify. An obvious example is where herbicide (weed killer) drifts on to unintended targets, causing twisted, discoloured foliage, sometimes killing all or part of the plant. Manure can be contaminated with herbicide from straw bedding and should always be sourced carefully where it is to be spread on the soil.

Pesticides and fertilizers can also harm plants when used incorrectly, so it is vital to follow precisely the instructions on all garden products (see p.27 and pp.46–49). Salt spray in coastal gardens or from treated roads can result in brown patches of dead foliage, while dog urine burns distinctive yellow circles on lawns where the grass has died back.

Applying fertilizer unevenly to a dry lawn can result in visible "burn".

This grape vine has been affected by weed killer spray carried by the wind.

CLIMATIC CONCERNS

When you are out tending to your garden on a mild spring day, it's easy to forget how hostile the weather can be. Plants grown outdoors are certain to be exposed to extreme conditions that can cause physical injury or weaken growth, both of which will spoil their appearance and increase their susceptibility to pests and diseases. Minimize problems by choosing plants to suit the conditions on your plot. Select fully hardy plants where winters are cold, plants with waxy grey (glaucous) foliage for dry, sunny spots, and bog plants for locations with persistently wet soil.

Hard frosts and high winds may scorch the leaves of perennials, such as bay.

FROST

When temperatures fall below 0°C (32°F) the watery content of plant cells can freeze, causing cell walls to rupture. Such frost damage is most often seen at the tips of evergreen shrubs, where leaves turn brown and stems die back. It is also common in autumn on tender plants, which turn soft and black from the tips and die back. The soil may also freeze (especially in pots), damaging roots that later rot as the weather warms.

Late frosts in spring nip new growth on deciduous shrubs and perennials, and can damage open blossom on fruit trees so that fruit fails to set. Wait until the risk of frost has completely passed before moving tender plants outdoors. These, as well as some vegetables, such as tomatoes and squashes, should be "hardened off" if raised under cover (see right).

HARDENING OFF

Plants that have been raised indoors or in a greenhouse – such as vegetable seedlings or half-hardy bedding – need to be exposed gradually to outdoor conditions before being planted out. This process of acclimatization, or "hardening off", takes about two weeks.

In the first week, move plants in their containers outdoors for 3–4 hours a day. Leave them in a sheltered, shady spot before bringing them back indoors at night. Gradually lengthen their daily outdoor exposure and move them into sunnier locations. In the second week, you can leave then outdoors overnight, as long as there's no danger of frost. Ensure that compost in their containers doesn't dry out. Plant out on a cloudy day and water well. Cover them at night with horticultural fleece in particularly cold or blustery weather.

TOP TIP IF YOU HAVE A LOT OF SEEDLINGS TO HARDEN OFF, PUT THE POTS IN A WHEELBARROW AND MOVE THIS IN AND OUT OF COVER.

Basil seedlings are a feast for slugs and snails; elevating them on a table while hardening off provides some protection from these pests.

Tomatoes subjected to high temperatures and bright sunlight in a greenhouse may display greenback, where the end of the fruit nearest the stem remains green.

High winds can shear off branches from trees and shrubs.

HIGH TEMPERATURES

Hot summer weather suits some plants well, but those with delicate foliage may develop dry, brown scorch marks if subjected to the heat of full sun. Extreme temperatures can be a particular problem in greenhouses, where adequate ventilation is essential and shading should be added to the glass in summer to prevent scorch. High temperatures may also cause leaves to wilt, sometimes because the soil is dry, and sometimes because water is lost through the leaves faster than it can be taken up through the plant's roots. This stresses plants, leaving them more vulnerable to pests, diseases, and disorders. Susceptible plants will grow poorly and fail to flower, and leafy vegetables may run to seed rapidly (bolt). It's sensible to keep plants well watered during unusually hot spells. Where high temperatures are the norm, grow species that are naturally adapted to cope with little or no irrigation.

WIND

Strong winds can break woody branches, flatten tall stems, and rip climbing plants from their supports. Plants with delicate foliage can also be scorched by the wind, which turns leaves brown at the edges and tips, particularly in spring when new leaves are still soft. Persistent wind can also exacerbate drought conditions by drying the soil and blowing away moist air that usually surrounds leaves.

RAINFALL

If young plants don't receive sufficient water, their growth will slow down and you can expect poor flowering and fruit set, or even plant death. If your garden has free-draining, sandy soil, you may need to water the plants every other day during periods of low rainfall.

Too much rain will also cause problems as nutrients – particularly nitrogen – may be washed (leached) out from light soil, resulting in deficiency (see p.14), or waterlogging in heavy clay soils. In saturated soils, roots quickly suffer from a lack of oxygen and foliage may begin to yellow. If conditions do not improve quickly, the roots will die back and fungi will move in to cause damaging rot in the roots and bases of stems (foot rot).

Drought-tolerant plants can make an attractive and resilient display.

TOP TIPS

- Choose plants that suit your local climate.
- Harden off young or new plants that have been kept under cover before planting out.
- Protect plants during winter.
- Create shelter from the wind and shade from summer sun.
- Avoid overfeeding to prevent excessive soft young growth that's easily damaged.
- Don't prune or trim too late, so that new growth has time to ripen before winter.
- Improve soil drainage to prevent wet soil over winter, which makes plants less resistant to cold weather.

HOW TO SPOT AN AILING PLANT

The sooner you detect that something is awry with a plant, the sooner you can start to diagnose the issue and determine what – if anything – you need to do to speed recovery, eliminate the problem, and stop it from spreading further around the garden. To spot problems reliably, you first have to familiarize yourself with the normal appearance of healthy plants.

Pests like these cabbage whitefly hide away over winter, so check the undersides of leaves in the spring.

Healthy leaves indicate that the roots are probably healthy too.

RECOGNIZING CHANGE

As you work in the garden throughout the year, make a point of noting the different habits, seasons of growth, foliage colour, and other characteristics of your plants. After a while, you'll be able to spot even subtle changes; for example, the dulling of what should be glossy leaves, upright stems looking floppy, or a delay in bud break. It is also worth making written notes of plants with unusual features that cause alarm, such as variegation, unusually shaped leaves, or new growth that is a different colour to mature foliage. Also, remember which plants are annuals and biennials, as they will not reappear year after year unless they have set their own seed.

PRIMARY SYMPTOMS

Primary symptoms are the damage caused directly by a pest or disease at the site where it attacks the plant; the holes eaten in leaves by insect pests, for example, or the distinctive raised orange pustules of rust fungi. They are the most helpful clues when trying to identify a problem.

Some of the most common primary symptoms are spots or marks on foliage, fuzzy or powdery fungal growth on any part of the plant, and galls or deformed leaves caused by feeding insects. However, by the time you spot these symptoms the damage might already have been done, particularly with pests such as codling moth and plum moth, which feed inside fruit, or soil-dwelling pests and diseases that affect roots. Losing crops or plants is frustrating, but once you know that a particular pathogen is present it will allow you to prevent or reduce the problem in future growing seasons.

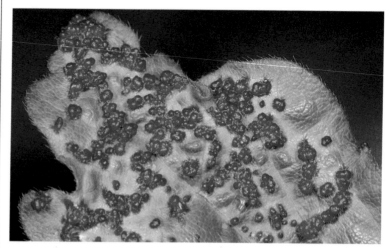

Tiny sap-sucking gall mites give away their presence by inducing areas of raised, abnormal growth.

SECONDARY SYMPTOMS

Plants under attack by pests or disease can display secondary symptoms, which affect the whole plant or parts that have not been directly infected. These symptoms provide a clear warning that the plant is stressed, and should prompt you to look for the cause. This may be an unseen problem beneath the soil, such as vine weevil larvae or clubroot, a bacterial canker, or a serious disease like honey fungus that you will want to prevent from spreading. Alternatively, your problems could simply be lack of water or nutrients that can easily be rectified.

Look out for common secondary symptoms, such as wilting, yellowing, or withered leaves, buds failing to open, dieback of shoots, or visible ooze. Sniff your plants to see if they give off unpleasant or mushroom-like smells, which may also indicate hidden problems. Always follow up your suspicions – don't assume the problem will just go away.

A wilted cabbage seedling could indicate drought or cabbage root fly.

KNOWING WHEN TO WORRY

Occasional leaf spots and low levels of pests on ornamental plants are usually of little concern and can be tolerated, but be on the alert for symptoms like wilting, yellowing, or browning leaves, or shoots starting to die back, which could indicate more serious problems.

Pest damage is a worry when it affects plants so that they can no longer serve the purpose for which they were planted. Hostas, for example, which are grown for their elegantly textural leaves, are of little value if shredded by slugs; similarly, aphid infestations that stunt the development of blossom on fruit trees need to be addressed urgently.

Time of year is a consideration: don't waste too much energy on pests and diseases like powdery mildew or slug damage at the end of the season, when many annual vegetables and perennials will soon die down, but be proactive if these problems occur in spring or summer. Finally, some harmless developments, such as the flattened growth caused by fasciation (see p.108), or leaf notches made by leaf-cutter bees, can even become welcome curiosities.

Wasps will feed on pears where birds have pierced the skin.

SECONDARY INFECTIONS

Any damage caused to plant tissue by a pest or disease becomes an easy target for new attacks, known as secondary infections. These can often be more serious than the initial problem and can mask it, making it tricky to diagnose how the trouble started. Secondary infections are often fungal diseases, such as grey mould, which enter the plant body via holes made by pests or splits in bark or fruits caused by disease or poor growing conditions. New pests can also move in to take advantage of holes in fruits, roots, or tubers. Avoid problems by pruning out injured growth or removing damaged fruit.

Significant dieback, such as on this gooseberry bush, warrants investigation.

Rolled rose leaves caused by rose sawfly activity are rarely a cause for concern.

YOUR PESTS AND DISEASES TOOLKIT

When it comes to making your garden flourish, no single method holds the key to success. Instead, savvy gardeners use an integrated approach, combining good husbandry with a range of preventative measures, barriers, natural controls, and treatments – sometimes including the use of chemicals – to raise resilient plants and protect them from pests and diseases.

A healthy and productive vegetable plot is the result of good cultivation and pest-control measures.

Pick off fading brassica leaves from maturing plants to stop them accumulating and encouraging pests and diseases.

KEEP PROBLEMS OUT OF YOUR GARDEN

Avoid bringing pests and diseases into your garden by buying plants and seeds from reputable nurseries and closely inspecting gifts from friends' gardens carefully before planting. Stop problems spreading and overwintering by keeping your garden weeded and tidy, as well as cleaning equipment and tools regularly. Plant pest- and disease-resistant varieties of vegetables and ornamental plants whenever they are available (see p.36).

KEEP PESTS AWAY FROM CROPS

Inexpensive physical barriers are an effective way to protect fruit and vegetable crops from pest damage. Netting, insect mesh, and horticultural fleece keep birds, egg-laying butterflies, and smaller insect pests away from plants (see p.34). Crop rotation prevents the build-up of pests and diseases in soil (see p.33) and should be used in tandem with barriers to prevent pests emerging from the soil beneath covered crops. Simple cardboard discs around brassica stems are enough to prevent cabbage root flies laying eggs successfully, but robust fencing is needed to exclude rabbits or deer from the garden.

Netting will prevent pests from reaching plants. Here cabbages are protected from butterflies and pigeons.

RAISE STRONG, RESILIENT PLANTS

Give plants an environment in which they can thrive and they will grow into robust specimens that are less susceptible to disease and more able to recover from pest damage. Many problems can be avoided by choosing plants carefully to suit your soil and climate (see p.24), improving the soil annually with plenty of organic matter (see p.26), and watering and feeding plants growing in containers and under cover regularly (see p.27).

Established plants usually need little input, so concentrate your efforts on caring for those at vulnerable stages of development, like delicate seedlings or recently planted perennials and shrubs that may need watering while their roots establish.

Add plenty of organic matter to the soil before planting.

WORK WITH NATURE

Welcome wildlife into your garden and a range of beneficial creatures will help to create a natural balance where pest numbers are kept under control by natural predators (see pp.28–29). Biological controls work in a similar way by introducing pest predators or parasites into the garden (see pp.42–43). Pesticides kill beneficial insects and upset this delicate equilibrium, which is one reason why many gardeners prefer not to use them.

Encourage lacewings into your garden: their larvae feed on aphids.

Purple basil growing alongside tomatoes helps to distract whitefly.

TRY DISTRACTION

Traps baited with beer, the scent of female insects, or the fragrance of a food plant (see p.35) can be used to lure pests and prevent their escape, often reducing numbers enough to limit damage to tender young shoots. Colourful and strongly scented companion plants (see p.35) have long been grown among crops to confuse pests, while sacrificial plants that are particularly attractive to pests are planted to draw them away from crop plants that are more valuable to the gardener.

CHEMICAL FIXES

Pesticides are marketed as instant solutions to pest and disease problems, but their effects are usually short-lived and many are harmful to beneficial insects. Increased regulation means that no products are available to treat some common problems, particularly on edible plants. Choose less toxic organic products wherever possible, because synthetic pesticides persist on plants for longer, making them more likely to affect beneficial creatures.

Spraying aphids will also kill any beneficial creatures that control their numbers, like this hoverfly larva.

Protective netting is a valuable asset in the kitchen garden, keeping hungry birds, egg-laying butterflies, and scratching cats away from fruit and vegetable crops.

PREVENTION AND TREATMENT

An integrated approach that combines a range of preventative and control measures is the best way to look after the health of your plants on your plot and the wider environment. Problems can often be prevented by cultivating strong plants, encouraging beneficial wildlife, and creating barriers against pests, but where issues do arise gardeners can also make use of many safe and effective treatments.

ALLOW YOUR PLANTS TO THRIVE

Every gardener wants their plants to flourish. In addition to looking good and cropping well, healthy plants are more resistant to attacks by pests and diseases and better able to shrug off any damage that does occur. Keeping your garden in good shape involves choosing plants to suit your plot, ensuring that their basic needs are met, and maintaining optimal growing conditions.

There are many plants that will thrive even in the shadiest parts of your garden.

PLANT BASICS

Plants need light, water, nutrients, and air to grow and remain in good health. Light is essential for photosynthesis, which produces energy-rich sugars for growth by harnessing sunlight in the green chlorophyll pigment present in leaves. Water is taken up from the soil by fine roots and evaporates from tiny pores in the foliage, drawing a steady supply upwards and preventing the plant from wilting. Nutrients from the soil dissolve in water, enabling plants to absorb them through their roots and transport them to wherever they are needed. Air moves in and out of plants through stomata (tiny pores in the leaves), providing carbon dioxide for photosynthesis. Selecting the right site, improving the soil, and feeding, watering, and pruning plants as necessary will help to ensure that these essential needs are fulfilled.

RIGHT PLANT, RIGHT PLACE

The simplest route to a healthy and easily maintained garden is to fill it with plants that are naturally suited to the local growing conditions. Variables including light and shade, rainfall, and seasonal temperature ranges, as well as aspect, drainage, and soil type and pH, can all factor into choosing the right plant for the right place. Bear in mind too that conditions can vary quite dramatically within your garden, where there may be several microclimates created by reflective walls, shade from trees, or the rain shadow of a fence.

Over many decades, garden plants have been collected from habitats throughout the world and bred to enhance their performance, so there are almost certain to be species and cultivars to suit your plot.

TOP TIP SEEK ADVICE ON WHAT TO PLANT WHERE FROM GOOD GARDENING BOOKS AND FROM THE WEBSITES AND STAFF OF SPECIALIST NURSERIES AND GARDEN CENTRES.

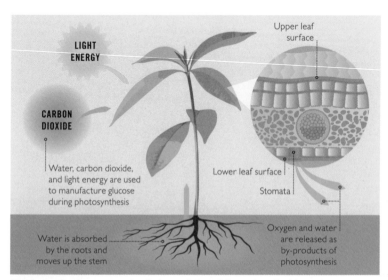

LIGHT ENERGY

CARBON DIOXIDE

Upper leaf surface

Lower leaf surface

Stomata

Water, carbon dioxide, and light energy are used to manufacture glucose during photosynthesis

Water is absorbed by the roots and moves up the stem

Oxygen and water are released as by-products of photosynthesis

Plants make the sugars that fuel growth from light, carbon dioxide, and water.

Weeding removes unwanted plants and is a perfect time for a ground-level check.

REGULAR MAINTENANCE

Once established in good soil, many plants need little attention other than watering in drought conditions. However, regular care will keep your plants in peak condition and make it easier to spot signs of trouble.

Weeding removes unwanted plants that compete for water, light, and nutrients, and that may also act as reservoirs of pests and diseases. Adding a thick layer of organic mulch – made from materials such as compost, leaves, and bark chippings – over the soil each autumn or spring improves growing conditions by suppressing weeds, adding slow-release nutrients, and keeping soil moist.

Regular pruning of trees and shrubs is needed to remove any dead or damaged stems that could provide an entry point for diseases, and to maintain an open branch structure that allows air to circulate. Lift, divide, and replant clumps of herbaceous perennials every three years so that weeds can be removed and any weak or diseased sections discarded.

PLANTS IN CONTAINERS

Plants in pots are entirely reliant on you to create and maintain good growing conditions. Ensure that the containers you use have adequate drainage holes in their base. Select the compost to suit the plant: peat-free multi-purpose compost is ideal for bedding and annual vegetables; soil-based compost suits perennials, shrubs, and trees; while ericaceous compost is needed to successfully grow lime-hating plants, such as rhododendrons, camellias, and azaleas.

Position the pots carefully to suit the plant – some grow well in shade, while others require full sun for at least part of the day. Container plants need watering frequently in the growing season – daily in the summer. Plastic pots retain water better than porous ones, and small pots dry out much faster

Broken crocks placed at the base of a container improve drainage.

than larger ones. All pot plants need feeding regularly with a liquid fertilizer. Reduce maintenance by installing a drip irrigation system and adding controlled release fertilizer to compost in spring.

SOIL TYPE AND PH

The physical properties of garden soil influence its ability to hold water and nutrients. Sandy "light" soils have coarse particles and are crumbly and free-draining; they warm up quickly but nutrients wash away easily. "Heavy" clay soils have fine particles and are sticky; they drain and warm up slowly but hold nutrients well.

A pH testing kit will show if your soil is acidic, alkaline, or neutral. Most plants thrive in neutral soils (pH 6.5–7), while lime-hating plants need acidic soils (pH 5–6). Alkaline soil (pH 7–8) reduces clubroot disease on plants in the cabbage family. Some nutrients become "locked up" in acidic or alkaline soils, leading to deficiencies. Lime can be added to increase pH, but it is difficult to reduce soil pH.

Heavy clay (left) contrasts with a light, sandy loam.

FEED THE SOIL

Plants take up all the water and nutrients that they need for growth from the soil, making this the most precious resource in your garden. Feeding and conditioning the soil is – not surprisingly – one of the best ways to produce healthy, productive plants that can resist and recover from pests and diseases. Regular applications of organic mulches or soil conditioners help to retain moisture, benefit soil-dwelling organisms, and add vital nutrients to the soil. Fertilizers are needed less frequently on healthy soils, but are always useful to boost performance and correct nutrient deficiencies.

When applying a compost mulch, be sure to leave a small gap around the base of plants.

Adding a thick mulch of garden compost to raised beds each year will help produce a bumper vegetable crop.

ADD ORGANIC MATERIAL

Mulch is a layer of material that is put on top of the soil around the base of a plant, where it helps block weed growth and retain moisture. Mulch can be made from a variety of materials, but only mulches with an organic base (see box, right) are broken down in the soil by beneficial creatures and microorganisms. Using such mulches releases nutrients that can be taken up by plant roots, and improves soil structure. Always use well-rotted organic mulch, because the initial stages of decomposition can actually withdraw valuable nitrogen. Apply mulch every year in autumn or early spring on heavy soils, but only in early spring on light soils (to prevent nutrients being washed away in winter), spreading it 2.5–5cm (1–2in) deep on the surface.

The same organic materials can also be used as soil conditioners, which are dug into the soil rather than laid on the surface. Use a fork to work organic matter into the soil at the ratio of one large bucketful per square metre (yard). Making your own compost (see *opposite*) will help keep costs down.

SOURCES OF ORGANIC MATTER
- Garden compost: free, convenient, but may contain weeds.
- Composted green waste: cheap, readily available, but may contain weeds.
- Spent mushroom compost: alkaline, so not for lime-hating plants; not for regular use unless aim is to increase pH.
- Leaf mould: free, fine texture, but slow to produce and may contain weeds.

MAKE A COMPOST HEAP

Composting garden waste and veg peelings from your kitchen is easy and produces a free supply of compost in 6–12 months. You can build your own compost bin to a simple design or buy one to suit your site. Place it in a warm spot with good access and keep the bin covered to exclude rain and weed seeds. Set the bin on soil rather than hard standing, to allow creatures essential for decomposition to move in.

Add an even mix of green nitrogen-rich waste (such as grass clippings, veg peelings, and leafy plant waste) and brown carbon-rich waste (fallen leaves, chipped wood, and even shredded paper and cardboard). Avoid including perennial weeds or cooked food.

Turn your compost periodically to aid even decomposition.

Place leaves in a net bag before steeping to make the tea easy to strain.

USE FERTILIZERS

Commercially available fertilizers in liquid or granular form can be useful in the garden. They can boost vegetable yields, help produce prolific flowers, and feed plants in containers. They usually contain all three main nutrients: nitrogen (N) for leafy, green growth; phosphorous (P) for root and shoot growth; and potassium (K) for flowering and fruiting. The relative amounts of the nutrients are shown on the packaging as an N:P:K ratio. Some fertilizers also supply micronutrients and trace elements.

Inorganic fertilizers (such as sulphate of ammonia and liquid tomato feed) are synthetic or mineral-based and make nutrients immediately available to plants. Organic fertilizers (such as bonemeal and poultry manure pellets) are derived from plant or animal sources and need to be broken down in the soil before nutrients can be absorbed.

Use fertilizers between spring and midsummer. Note that excessive or late feeding can scorch leaves or produce soft growth that is vulnerable to pests, diseases, and frost damage.

Granular fertilizer is convenient to use and is typically spread in spring.

HOMEMADE FERTILIZERS

Organic liquid fertilizers are easy to make and can be a valuable source of nutrients. "Compost teas" can be made simply by steeping the leaves of potash-rich comfrey or borage, or nitrogen-rich nettles, in water. The liquid from wormeries is also an excellent general fertilizer.

To make compost tea, cut leaves from your chosen plant, pack them into a plastic bucket, and weigh them down with a brick. Add 10 litres (18 pints) of water for every 1kg (2¼lb) of leaves, and cover with a lid to reduce odours. Leave nettles to steep for around two weeks and comfrey and borage for six weeks. Strain off the brown liquid and dilute one part tea with ten parts water before watering around plants.

RECOGNIZING BENEFICIAL WILDLIFE

Every garden hosts a diversity of wild creatures, only a small minority of which are pests. Many more are a gardener's allies, because they pollinate flowers, prey on pests, and help to improve the quality of the soil. Learning to identify beneficial creatures will not only allow you to protect them from harm, but will give you insight into the fascinating ecosystem of your own back yard.

Painted lady butterflies are prolific pollinators, and are the most widespread butterflies in the world.

Planting fennel will attract beneficial hoverflies into your garden.

HELPFUL INSECTS

Insects are essential for a healthy and productive garden. Pollinators transfer pollen between flowers as they buzz, crawl, or flutter around, enabling edible plants to produce good crops of pods and fruit, and ornamentals to set berries and seed for the following year. Hardworking honeybees and bumblebees usually get the credit, but hoverflies, butterflies, moths, beetles and many other insects also move between blooms, doing the same job.

A surprising number of insects prey on garden pests. Aphids are a favourite for adult and larval ladybirds, lacewings, earwigs, hoverflies, and some midges. Solitary wasps eat many kinds of insect, while ground and devil's coach horse beetles also have a varied diet that extends to slugs. Soldier bugs will eat caterpillars and beetle larvae.

Parasitic wasps and some flies lay their eggs on the eggs, larvae, and pupae of a range of insects, so that their own larvae can – rather gruesomely – feed within them before emerging as adults.

MAMMALS AND AMPHIBIANS

Often nocturnal and reclusive, garden mammals and amphibians are relatively large animals with appetites to match, meaning that just a few of them can keep populations of common pests in check. Hedgehogs and shrews use their sensitive noses to seek out slugs, worms, and all kinds of insects. Moles dig tunnels that help to aerate soil and improve drainage and consume insect larvae, such as chafer grubs – activities that weigh against their annoying habit of leaving mounds of soil on lawns or in cultivated beds. Frogs and toads should also be encouraged into the garden, as they feed on slugs and other pests.

Larger frogs will prey on insects like flies, moths, and grasshoppers.

BIRDS

Many of us welcome birds into our gardens for their beauty and song, but these creatures also play a valuable role in controlling pests. Robins will follow you round the garden as you dig, picking out juicy insect larvae from the disturbed soil. Blue tits, coal tits, and great tits descend in groups to pick caterpillars, aphids, and beetles from trees and shrubs, and are agile enough to pick overwintering eggs and insects from tiny fissures in the bark of even the highest branches. Song thrushes prey on insect larvae, and are also skilled at breaking snail shells against a favourite flat anvil stone to eat the defenceless creatures inside.

Broken snail shells around an anvil stone indicate songthrush feeding.

Robins feed on a variety of insects, but you can use seeds or crushed peanuts to attract them into your garden.

SOIL-DWELLERS

Healthy garden soil is teeming with life. Many of the beneficial creatures living on or beneath its surface are known as decomposers. They feed on dead plant and animal material and break it down to release nutrients back into the soil and improve its structure. Long, pale pink earthworms tunnel through soil, improving aeration and drainage, and pull fallen leaves and other detritus down into the soil from the surface to feed on. Smaller, red brandling (or tiger) worms are efficient decomposers, and are often found on compost heaps or other areas with lots of decaying matter.

Long, many-legged millipedes, with their domed, black bodies, and grey woodlice, with seven pairs of legs, are both often found on the soil surface, helping to break down decaying wood or dead plant material. Orange-brown centipedes are carnivorous, preying on insects and their eggs both on and in the soil, and so controlling soil pests.

TOP TIP ATTRACT EARTHWORMS INTO YOUR GARDEN BY ADDING ROTTING ORGANIC MATERIAL SUCH AS MANURE OR COMPOST TO YOUR SOIL, OR BY BURYING DRY FALLEN LEAVES. KEEPING YOUR LAWN WELL WATERED WILL ALSO ENCOURAGE WORM ACTIVITY.

Earthworms are every gardener's friend – treat them as unpaid labour!

ENCOURAGE YOUR ALLIES

Collectively, gardens are a patchwork of valuable habitats with the potential to support a diversity of wild creatures, many of which have a positive effect on the health and productivity of your plot (see pp.28–29). It makes sound sense to manage your garden in a way that attracts more of these animals. Even on the smallest plot, setting out some food and water will help, as will allowing parts of your garden to go unmown and untidied to provide refuges for small mammals and invertebrates. Limiting your use of pesticides is desirable too, as these chemicals may kill beneficial creatures.

PROVIDE AMPLE FOOD AND WATER

The surest way to tempt beneficial wildlife into your garden is to provide plentiful food and fresh water. Hang bird feeders filled with diverse seeds, nuts, and fat balls (so that you attract a range of species) in trees or sturdy shrubs, and fill a wide, shallow bird bath or dish with fresh water for them to drink and bathe in. Remember to replenish supplies regularly because birds will return day after day to a known, reliable source.

Pack borders with a variety of nectar- and pollen-rich flowers. This will draw in pollinating insects, including hoverflies whose larvae feed on aphids. Native wildflowers are often key to insect life cycles, but cultivated plants, especially those with heads made up of multiple flowers, are also excellent food sources for pollinators (see box, right) and fit easily into existing displays.

Building a pond – even a small one – will quickly attract frogs and toads which devour flies, slugs, and snails. Add gently sloping sides; these allow mammals and insects to visit for a drink and escape if they slip into the water.

A peanut feeder brings tits into the garden: these birds will also happily pick aphids from your plants.

BOOST YOUR BOUNDARIES

Bare fences and walls are of little use to wildlife, but clothed in flowering and berry-bearing climbers and shrubs, they suddenly become a haven where birds, insects, and other invertebrates can find food and shelter. These vertical spaces are especially valuable in small gardens, where space for planting is limited. Boundaries lined with trees, shrubs, or hedges have even greater potential as sources of food and cover, also providing nesting sites for birds and vital corridors along which wildlife can move safely between neighbouring gardens. Where possible, don't block access between gardens to allow hedgehogs to forage more widely.

The autumn berries and white spring flowers of firethorn (*Pyracantha*) are a feast for wildlife.

Leave grasses standing over winter to provide cover for beneficial wildlife.

PROVIDE WINTER COVER

At the end of the growing season, leave clusters of stems of herbaceous plants and grasses standing, rather than cutting them down. They will provide winter shelter for invertebrates, amphibians, and small mammals, which will then be ready in spring to feed on newly emerging pests.

Fallen leaves should be raked up from lawns and borders, but left at the base of hedges, where they make a perfect place for hedgehogs to hibernate. A log pile in a quiet corner of the garden will attract decomposers year-round and also provides welcome winter shelter for frogs, toads, and small mammals.

CHEMICAL CAUTIONS

Chemical controls are a highly effective way to kill insect pests, but they can also kill beneficial insects. Even if the population of predators isn't directly harmed by pesticides, using chemicals will entirely wipe out their local food source. This means that there will be fewer predators to limit pest numbers when they inevitably do recover and so encourages a long-term reliance on chemical controls. Many gardeners choose not to use pesticides to avoid upsetting this natural balance. Alternative methods are available to control pests and protect plants (see *pp.20–21*) and pesticides should only be used where absolutely necessary.

Predatory insects like ladybirds are likely to be harmed by pesticide use.

TOP TIP BEE HOTELS SECURED TO SUNNY, SOUTH-FACING WALLS PROVIDE NESTING SITES FOR MANY SPECIES OF SOLITARY BEE, WHICH ARE EXCELLENT POLLINATORS.

PLANTS FOR POLLINATORS

To encourage pollinators into your garden, include the following in your plantings, siting them in a sheltered sunny spot if you can.

SPRING Crocuses • Alliums • *Pulmonaria* • Apple, pear, and cherry blossom

SUMMER *Veronicastrum* • Lavender, thyme, and other Mediterranean herbs • *Buddleja davidii* • *Cistus* • *Achillea* • Fennel

AUTUMN *Hylotelephium* • *Agastache* • Michaelmas daisies

WINTER *Mahonia* • Hellebores • Snowdrops

PREVENT PROBLEMS

There are many measures that you can take to keep pests and diseases away from your plot. This is far preferable to fighting a rearguard action, trying to eradicate established diseases or to control booming pest populations. Simple precautions, such as cleaning tools and pots, will cut through infection cycles and prevent pests and diseases overwintering in your garden, denying them an easy start in the spring. Being alert for pests and diseases when introducing new plants into your garden, and regular health checks, especially while seedlings are establishing, will pay dividends.

Reject any bulbs with brown patches, softening, or mould as they may harbour disease.

CLEAN AND TIDY

Good garden hygiene will help prevent the spread of pests and diseases. Always rake up fallen leaves and fruits that show any signs of infection promptly to prevent fungal resting bodies or insect larvae entering the soil to overwinter. Don't add such waste to your compost heap, but burn it or dispose of it in a green waste bin. Wash pots and seed trays with hot soapy water before use and clean the inside of your greenhouse, including any staging, with hot soapy water or disinfectant. Wash all your pruning tools after use and sharpen them regularly so that they make clean, fast-healing cuts. Avoid accidental damage to plants by using separate, clearly labelled, equipment for watering and the application of pesticides and fertilizers.

KEEP PROBLEMS OUT

Some pests and diseases are tenacious. Once they get a grip on your plot, they can be hard to eradicate and may prevent you growing certain groups of plants for many years. Avoid bringing problems into your garden in the first instance. Buy plants, bulbs, and seeds from reliable sources and choose fruit bushes and seed potatoes that are certified disease-free. Don't accept any plants showing symptoms like leaf spots or discoloured foliage, and, if possible, check roots for any signs of pests, disease, or rotting. Inspect bulbs (ornamental, onion sets, and garlic) carefully before planting to ensure that they are firm, with no signs of fungal disease.

It's great to share plants with friends and relatives, but check plants or bulbs for problems as your guard may be down. Reduce the risk to established plants, such as box, by quarantining new introductions in containers for four weeks before planting.

Clean your secateurs with household disinfectant after each use.

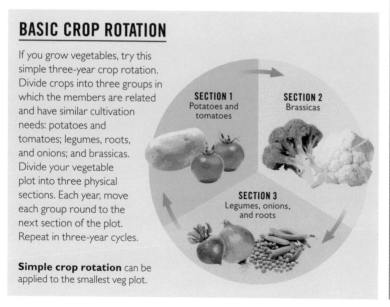

Organizing related crop groups in several raised beds makes it easy to plan a crop rotation.

DON'T STORE UP TROUBLE

Pests and diseases tend to accumulate in the soil around long-established plants or where related bedding plants or vegetable crops are grown year after year. This can lead to serious outbreaks or just a gradual decline in plant performance. Similar problems can occur when plants are grown in containers if the compost is not replaced with each new planting. If you suspect that your soil may be harbouring pests or diseases, try to identify the agent responsible, and avoid replanting anything from the same plant family at the site, because it is likely that the new plants will succumb quickly to the same problem.

Gardeners traditionally use crop rotation (see box, below) to avoid the build-up of pests, diseases, and nutrient deficiencies in the vegetable plot. It is an effective way to raise healthy crops, even in small gardens.

TOP TIP AVOID SPREADING PATHOGENS IN CONTAMINATED SOIL BY CLEANING BOOTS AND TOOLS THOROUGHLY AFTER WORKING IN AFFECTED AREAS.

BE VIGILANT

Problems with pests, diseases, and disorders are always easier to deal with when caught early. It pays to get out into the garden as often as possible during spring and summer to keep an eye on early growth and check for common symptoms (see pp.18–19).

Pull up weeds as soon as they appear to prevent them seeding around your borders and providing refuges for pests and diseases, and manually pick off any early-arriving pests and young leaves showing signs of disease. Squash tiny eggs on foliage, newly hatched caterpillars, and any aphids at shoot tips; this is often enough to keep numbers under control and to delay, or even stop, their spread.

Regular jobs, such as weeding and deadheading, give you an opportunity to check for signs of pests and diseases.

BASIC CROP ROTATION

If you grow vegetables, try this simple three-year crop rotation. Divide crops into three groups in which the members are related and have similar cultivation needs: potatoes and tomatoes; legumes, roots, and onions; and brassicas. Divide your vegetable plot into three physical sections. Each year, move each group round to the next section of the plot. Repeat in three-year cycles.

SECTION 1
Potatoes and tomatoes

SECTION 2
Brassicas

SECTION 3
Legumes, onions, and roots

Simple crop rotation can be applied to the smallest veg plot.

KEEP PESTS AWAY

Entirely excluding pests from your garden is impossible, so the best way to prevent them from causing damage is by using barriers to deny them access to your plants. These defences can take the form of nets, meshes, tunnels, and cloches, traps that collect or kill the miscreants, or deterrents that make your plot a hostile place for pests. Such physical measures reduce or remove the need for pesticides, and leave populations of beneficial insects unharmed. Many gardeners rely on these methods to produce healthy fruit and vegetable crops.

Simple plastic or card collars that fit round the base of plants are enough to protect brassicas from cabbage root fly.

PHYSICAL BARRIERS

Keeping plants and pests apart is an effective way to prevent damage. Sturdy perimeter fencing is needed for large and destructive mammals: it must be 1.2m (4ft) high and buried 30cm (12in) deep to keep out rabbits, and at least 2m (6ft) high to exclude deer. You can enclose newly planted trees in special tree tubes to protect from rabbits and deer. These tubes also aid early growth by creating a mild microclimate for the growing plant.

Barriers of netting, mesh, or fleece keep birds, egg-laying butterflies, carrot flies, and other pests off fruit and vegetable crops. Select the right mesh size to exclude the relevant pests (see table, below). Keep the netting clear of foliage, supporting it on frames made of wood, aluminium tubing, or hoops of thick wire or alkathene pipe. Rotate crops annually (see p.33) to stop pests that overwinter in soil from having a ready food supply the following spring.

Collars around the bases of young brassicas prevent cabbage root flies laying eggs. Barriers of grease or tree glue around the trunks of fruit trees stop flightless female winter moths climbing into the branches to lay eggs.

Mesh protects strawberries from the attentions of birds. Make sure to elevate the netting well above the plants.

PEST TYPE	MESH SIZE
Birds eating soft fruit	20mm (⅘in)
Wood pigeons, cabbage butterflies, cats scratching seedbeds	7mm (¼in)
Carrot flies, flea beetles, aphids, whitefly, wasps	Approx 1mm (½₅in) mesh or fleece

TOP TIP SECURE NETTING AT THE BASE WITH PEGS OR BRICKS TO PREVENT IT BLOWING AWAY AND BIRDS BECOMING ENTANGLED.

Beer traps hung in fruit trees reduce crop damage by wasps.

TRAPS

Traps lure pests away from vulnerable plants, but are also useful for monitoring pest numbers in spring, alerting you to the need for other measures.

Jam jars half-filled with beer attract slugs and snails when buried just above soil level, and trap wasps when hung in fruit trees and covered with lids pierced with a central hole. Traps laced with the scent of raspberry plants are available to control raspberry beetle. Wireworms burrow into old potatoes or carrots buried in the soil, which can then be unearthed and disposed of, while earwigs can be caught in plant pots that have been filled with straw and set upside-down on canes.

You can detect the presence of pests, such as whitefly, in your greenhouse by hanging sticky traps, but you'll need to use other methods to control them. Likewise, pheromone traps, which use the scent of female codling, tortrix, or box moths to attract and trap males, help monitor numbers so you can time any necessary treatment accurately.

COMPANION AND SACRIFICIAL PLANTING

Traditionally, companion planting uses the strong scents of certain aromatic species to deter pests; for example, growing garlic chives near carrots to ward off carrot fly, French marigolds (*Tagetes*) to distract whitefly from tomatoes, or sage and thyme around brassicas or roses to confuse aphids. It's likely that the benefits of companion planting owe much to the mixture of plants, which makes it harder for pests to locate their target; moreover, flowering companion plants attract a variety of beneficial insects, like hoverflies, to prey on pests. French marigolds also secrete a chemical from their roots that deters eelworms.

Sacrificial plants sited near vulnerable crops may draw some pests away and reduce, but not prevent, damage. Try using nasturtiums to tempt blackfly away from beans and butterflies away from brassicas, and sacrificial lettuces to distract slugs and snails from other crops.

Nasturtiums are here grown as a sacrificial crop alongside broad beans.

DETERRENTS

Deterrents can scare away unwelcome wildlife and pets, but need to be changed regularly because creatures will become accustomed to them. Numerous odour and irritant repellents are available for cats, foxes, and deer, but they must be reapplied after rain. A cat repellent plant, *Plectranthus ornatus*, often sold as "Scaredy Cat", may deter pets but is often just ignored. Electronic scarers that emit sound or spray water are effective for cats and foxes, and vibrating devices can deter moles, but need repositioning periodically. Wood pigeons avoid gardens where a kite in the shape of a bird of prey remains airborne. Humming tapes, tin foil, or CDs hanging from cotton above crops may also keep birds away until they get used to them.

Solar-powered animal deterrents work by emitting high-frequency sounds.

LOOK FOR ALTERNATIVES

Some plants in your garden may be beset by problems year after year. In such cases – and especially with annual plants – it may be more sensible to sidestep the pests or diseases responsible than to vainly tackle them head-on. Try finding alternative species that have been bred with disease resistance, or that are just less susceptible to damage. Another approach is to experiment with timing your sowing or planting, or tweaking your cultivation methods, so that you provide protection for your plants at vulnerable growth stages or when pests are more active.

Varieties of New Guinea *Impatiens* offer resistance to downy mildew.

CHOOSE PLANTS WITH RESISTANCE

Choosing plants that have been bred with pest and disease resistance reduces the need for other control methods and may be your only option where no chemical controls are available. Fruit and veg varieties are marketed on their ability to fend off pathogens like clubroot, potato blight, or powdery mildew; others may be resistant to bolting.

Look for ornamental and fruit trees, or veg plants like tomatoes, which have been grafted on to disease-resistant rootstocks. Some species or cultivated varieties of ornamentals, including roses, clematis, and busy lizzies (*Impatiens*) have natural or bred resistance to pathogens, including downy mildew, clematis wilt, and rose black spot.

FIND ALTERNATIVES

Rather than struggle to grow plants that are not thriving, look for near alternatives that are more likely to succeed. Where specific vegetable crops are being ruined by pests or diseases, browse the websites or catalogues of seed merchants for less susceptible options (see *table, right*).

If traditional fruit crops are struggling, try less common, trouble-free crops, such as goji berries, honeyberries, or chokeberries (*Aronia*). And if your ornamental plants are affected by a pest or disease, there are plenty of less vulnerable alternatives: for example, box plagued by box blight or box moth can be replaced with yew or *Berberis darwinii* 'Compacta', and numerous plants,

Asparagus peas are a pest- and disease-resistant alternative to peas.

including peonies, hellebores, and buddlejas are unpalatable to slugs and snails, rabbits, and deer. Select hardy plants where winters are cold, and avoid planting lime-hating azaleas and camellias where the soil pH is unsuitable.

PROBLEM VEG CROP	ALTERNATIVES
Hearting lettuce	Loose-leaf lettuce, mizuna, rocket
Cabbage	Kale
Spinach	Callaloo, chard, New Zealand spinach, perpetual spinach
Tomatoes	Tomatillos
Maincrop potatoes	First early potatoes
Peas	Asparagus peas, broad beans
Calabrese	Broccoli raab

Beetroot may bolt, but you can control this tendency by waiting for soil to warm before sowing seeds in spring.

CHANGE YOUR TIMING

Experienced gardeners know that timing is everything, particularly in the vegetable plot. Wait for the soil to warm before sowing outdoors in spring, because cold seeds and seedlings develop slowly, making them easy prey for pests and diseases. Sowing root crops in cold conditions, or leafy salads in hot summer weather, can also cause them to bolt and run to seed early. Sowing and harvesting can be timed to miss periods when crops are vulnerable to a particular pest or disease; peas sown in early spring, for example, should escape pea moths, while early potatoes harvested in June avoid potato blight. Successional sowings, where small quantities are sown every two or three weeks, will help you determine the best time to sow. They will also be harvested faster than large gluts, which remain in the soil longer, making them more vulnerable to attack.

> **TOP TIP** 'EARLY' AND 'DWARF' VARIETIES OF VEGETABLES TEND TO MATURE FASTER, MAKING THEM LESS LIKELY TO SUCCUMB TO PROBLEMS.

TRY CONTAINERS OR GROWING UNDER COVER

When you plant directly in the soil, you surrender some control over your plants' environment. Using containers or raised beds, however, lets you choose the soil conditions suited to your plants' needs, and allows you to renew the soil if you suspect that it harbours any pests or diseases. Plants grown in containers suffer less from attack by slugs and snails, and can also be moved under cover to provide protection from cold and wind.

Growing plants under cover enhances your control over the environment. Young vegetables or ornamental plants can be sown earlier in a greenhouse or conservatory, or on a window sill, giving them a head start in cold climates, and producing sturdy young plants that will grow away quickly when planted outdoors. Growing under cover also offers protection from some diseases, such as tomato blight.

Citrus trees make attractive specimens in pots, and can easily be moved under cover for winter.

PLANTS UNDER COVER

Cultivating plants in the warm, sheltered conditions within a greenhouse or on a windowsill is the ideal way to protect them from harsh climatic conditions. However, growing under cover is not without its own challenges, because the plants rely entirely on you for all of their needs: you must provide them with water and nutrients, and manage the temperature to avoid extremes of hot and cold. Pests and fungal diseases will flourish in the warmth and humidity under cover, but the enclosed environment of a greenhouse gives you additional options for their control.

Space is at a premium in a greenhouse, but don't be tempted to crowd your plants.

REASONS TO GROW UNDER COVER

Raising plants in a greenhouse or cold frame, or on a warm, bright windowsill indoors, allows you to protect them from the worst of the elements and to grow an exciting range of edible crops and ornamental plants that would not thrive outdoors. Tender ornamentals can often be overwintered under cover and moved outside for summer once the risk of frost has passed. Sheltered conditions are ideal for raising seedlings and crops, such as figs and chillies, that are unlikely to ripen outdoors in cooler regions. Growing under cover also provides some protection from diseases that infect wet foliage, such as tomato blight and peach leaf curl.

Ventilate your greenhouse on warm days to reduce the temperature within.

POTENTIAL PROBLEMS

Pests such as aphids, whitefly, and red spider mites love the warmth under glass. They are active for longer and produce more offspring than they do outdoors, and are protected from their natural predators. This means that pest numbers can increase very rapidly.

Cramped growing conditions and poor ventilation can also result in a lack of air circulation and high humidity, which both encourage fungal diseases like damping off and grey mould. Soft spring growth is particularly prone to infection, especially if it has already been scorched by intense sunlight.

As there is no rainfall under cover, plants need to be watered frequently, which may mean twice daily during hot summer weather, and fed regularly too. Pollinating insects can also be in short supply, which can lead to poor fruit set.

Strawberries grown under cover may need hand pollinating because natural pollinators are in low numbers.

HEALTHY GREENHOUSE GROWING CONDITIONS

Practising good hygiene will stop pests and diseases spreading, and creating consistent growing conditions will encourage plants to thrive. Clean greenhouse glass and staging annually with hot, soapy water. Do the same with your pots, seed trays, and propagators. Remove any dead or diseased plant material regularly.

Install a max/min thermometer to record daily temperature extremes. This will alert you to conditions that are unsuitable for particular plants, so that you can regulate temperatures by opening doors and vents. Apply shading paint or fabric to glass in summer to stop intense sunlight from scorching foliage and fruit, and use a heater to guarantee frost-free nights for tender plants during winter. Space plants adequately and evenly to allow a good circulation of air, which will help to prevent fungal infections.

Applying shading paint to the glass of a greenhouse in summer helps prevent scorch and lowers the temperature.

EFFECTIVE SOLUTIONS

You should ideally check the growing conditions in your greenhouse twice a day in spring and summer, or install automatic vent openers and irrigation to maintain optimal moisture and temperature. On hot summer days you can cool your greenhouse by damping down its central path with water.

A few simple measures will help control pest numbers under cover. Hang yellow sticky traps above your plants to catch pests such as whitefly, thrips, leaf miners, and greenfly. Grow companion plants, such as French marigolds, to deter insects (see p.35), and reduce problems with red spider mites by misting plants to increase humidity. Biological controls (see pp.42–43) are effective in enclosed spaces, and safe to use with edible crops. Chemical controls can be used on plants listed on their label, but should not be used where biological controls have been released. Fumigants can be an effective way to treat an entire greenhouse affected by insect pests.

Insects are attracted to the yellow colour of sticky traps. Check the traps to see what is present in the greenhouse.

TACKLING PROBLEMS

When a problem arises, the first step towards dealing with it effectively is a correct diagnosis. This book provides two straightforward routes to help you identify pests, diseases, and disorders quickly, along with all the advice you need to tackle them with confidence. Once you know what you are dealing with, taking action as soon as possible is likely to increase your choice of potential control methods and your chances of resolving or controlling the issue successfully. A rapid response also helps to minimize the impact on the plant so that it can be nurtured back to health.

Pests don't make themselves easy to spot. Always check the undersides of leaves.

IDENTIFY THE PROBLEM

Distinguishing between different pests, diseases, and disorders can be tricky for new gardeners. First you need to spot the symptom. Pests may be directly visible or produce telltale signs like sticky honeydew, distorted shoots, trails or marks. Diseases may cause spots or brown patches or produce mould or oozing wounds. See pp.18–19 for fuller guidance on how to spot ailing plants. Once symptoms are spotted, you can use this book to reach a diagnosis.

Plants of one group – such as shrubs, bulbs, or roses – tend to be prone to similar issues. Start by finding the entry for the group (see pp.50–75). Each contains a table that lists the symptoms that may affect each part of the plant – its leaves, flowers, and so on. Once you have found a description that best matches the observed symptom, follow the page numbers given in "Possible Causes" to read more about likely culprit(s) in the Directory of Pests, Diseases, and Disorders (see pp.76–141).

The Directory is organized by the part of the plant that displays symptoms, so you can also just leaf through the relevant pages until you find a photo and description that corresponds to the observed symptoms.

ACT PROMPTLY

Taking action as soon as symptoms appear minimizes damage, allows plants to recover faster, and makes it possible to manage problems without chemicals.

Check leaves and shoot tips weekly and pick off aphids, caterpillars, and other pests by hand. This can prevent infestations, but you must persist as insects reproduce quickly – one aphid may produce five offspring per day for up to 30 days. Where foliage has been chewed, hunt for slugs, snails, lily beetles, and vine weevils and remove any culprits hiding close by. If the stems of outdoor-sown seedlings are severed, sift through surrounding soil to find the cutworms or chafer grubs responsible.

Pick off leaves, flowers, or fruits at the first sign of disease to prevent symptoms spreading. Biological controls (see pp.42–43) are best applied before pest numbers are too high and it also makes sense to use chemical controls (see p.44–47) as early as possible.

Pick off caterpillars to prevent them devouring foliage.

Remove leaves showing early signs of disease.

NURSE PLANTS BACK TO HEALTH

Plants affected by pests and diseases will often recover well if given extra care and improved growing conditions. You can rectify foliage discolouration and fruit damage associated with nutrient deficiencies, like nitrogen deficiency or blossom end rot, by applying a suitable fertilizer, improving soil with organic matter, and watering plants more consistently. Similarly, regular watering and feeding often help plants to cope with minor damage caused by common pests and diseases, such as aphids, capsid bugs, and rusts, and will boost healthy new growth after an attack. Plants stressed by drought are more susceptible to fungal diseases, like powdery mildew, and may recover with regular watering and the addition of organic matter to the soil.

Feed and water tomato plants regularly where nutrient deficiency or blossom end rot have been a problem.

PREVENT TROUBLE SPREADING

Removing growth affected by many diseases, including fireblight and clematis wilt, can halt the spread of symptoms and may save a plant. Cutting away damaged tissue also helps prevent secondary infections taking hold and causing further problems.

Sometimes, there is very little you can do to treat a pest or disease, particularly if it has affected the roots or has caused serious symptoms before it was diagnosed. In such cases, remove the plant quickly, along with its roots and surrounding soil if necessary, to stop problems like honey fungus, potato blackleg, or tulip fire spreading. Plants infected with less serious fungal diseases (such as powdery mildew, grey mould, or rust) can be added to home compost heaps, but most pathogens will survive their relatively low temperatures, so other diseased material should be sent to council green waste sites or burned, where bonfires are permitted.

Rake up and remove any fallen leaves and other plant material that could carry pathogens and dispose of it carefully to avoid spreading infection.

BIOLOGICAL CONTROL

Plant pests all have natural predators and there's a lot you can do to attract these helpful creatures into your garden (*see pp.30–31*). The practice of biological control goes a step further by giving you the means to introduce beneficial organisms in a more targeted way and so reduce the populations of pests that are causing problems. These pest enemies are available by mail order, are easy to apply, and are an environmentally friendly alternative to pesticides or for use where chemical controls are unavailable. Biological controls are particularly effective against a range of greenhouse pests.

HOW BIOLOGICAL CONTROLS WORK

Biological control is the deliberate targeted introduction of pest predators, parasites, and nematode worms capable of infecting pests with disease. It does not strive to wipe out whole populations of pests, but to reduce their numbers to a low level that can be tolerated by plants.

Predatory insects and mites prey on pests during at least one stage of their life cycle, while parasitic insects lay their eggs inside pests, so their larvae feed within the pest's body before emerging as adults. Biological controls based on these creatures are most effective in the confines of a greenhouse, where pests are concentrated and easily located, the predators tend not to disperse, and the consistently warm conditions allow predators to multiply rapidly.

Most of the biological controls that are available for controlling outdoor pests, such as slugs and vine weevils, are pathogenic nematodes. These microscopic worms are watered on to foliage and soil, where they enter the target pests' bodies and infect them with bacteria that cause them to die.

> **TOP TIP** CHOOSE BIOLOGICAL CONTROLS IF YOU WANT TO AVOID OR REDUCE THE USE OF PESTICIDES ON YOUR PLOT.

Biological controls are ideal for use on edible greenhouse crops.

REASONS TO CHOOSE BIOLOGICAL CONTROLS

Biological controls have many advantages over chemical pesticides. Each agent targets a specific pest, or pest group, and will not harm other wildlife, plants, or people. They are safe to use on plants which may be damaged by pesticides and on edible crops where chemical controls are unavailable.

Using natural predators prevents pest species from developing resistance to pesticides and leaves no chemical residue on plants. When used correctly, most biological controls will remain effective for as long as there are pests present for them to feed on, so providing persistent control.

The parasitic wasp *Encarsia formosa* is a biological control agent that lays its eggs on the larval scales of glasshouse whiteflies.

The predatory mite *Phytoseiulus* is used to kill red spider mites. Each one can eat up to 20 larvae per day.

The parasitic wasp *Encarsia formosa* is easily introduced into the greenhouse on a card that is hung from plant stems.

EFFECTIVE APPLICATION

Biological controls cannot be used preventatively, because the pest has to be present for them to work. You can get an early warning of the presence of pests by hanging sticky yellow traps in your greenhouse (see p.35). As soon as the pests appear, it's time to apply biological controls before their numbers have a chance to build up.

Make sure that you have correctly identified the pest before choosing a suitable biological control product.

Follow the manufacturer's instructions precisely and check that the conditions are right for the predatory creature to thrive. For example, each product will have a minimum temperature at which it will be effective, and nematode agents may need specific soil conditions. In all but the coldest climates, controls can be applied from spring to early autumn, which is when plants are in growth and pests will be active. Some products, like nematode slug controls, are most effective when applied several times during the growing season.

PEST	BIOLOGICAL CONTROL
Aphids (greenhouse)	*Aphidoletes aphidimyza, Aphidius* sp., *Adalia bipunctata*
Red spider mite (greenhouse)	*Phytoseiulus persimilis, Amblyseius* sp.
Whitefly (greenhouse)	*Encarsia formosa*
Vine weevils	*Steinernema* sp., *Heterorhabditis* sp.
Slugs	*Phasmarhabditis hermaphrodita*
Box tree moth caterpillars	*Steinernema carpocapsae*

Two-spotted ladybirds will continue to multiply and eat aphids throughout the growing season.

BIOLOGICAL CONTROLS AND PESTICIDE USE

Pesticides may harm organisms that are biological control agents, so the two forms of pest control should never be used in tandem. The use of pesticides should also be avoided if you are planning to use biological controls in the future, because active ingredients in synthetic pesticides (see p.47) can remain active on plants at levels high enough to cause harm to insects for up to 10 weeks. While the persistence of organic pesticides is usually much shorter, they can also present a risk to beneficial insects.

Biological controls are an alternative to chemical treatments and must be given time to reduce and maintain pest populations at tolerable levels without any intervention with pesticides to upset the balance that has been created.

UNDERSTANDING CHEMICAL CONTROLS

Pesticides are chemicals that eliminate animal pests or fungal diseases. Their rapid action is appealing, but the protection they provide is usually short-lived. What's more, if they drift away from their targets or enter water courses, they can pollute the wider environment. Understanding how pesticides work, and how to use them effectively and safely, will help you decide if they are right for your garden.

Pesticides may be solid pellets, powders, liquids, or fumigants.

When using contact pesticides, be sure to spray the underside of leaves

REACHING THE TARGET

Chemicals known as contact pesticides must make direct contact with pests in order to work, so have to be applied over the whole plant, including the undersides of leaves. In contrast, systemic pesticides are absorbed by the plant and carried through its sap to untreated areas, reaching pests and diseases hidden among curled leaves. Other pesticides are mixed with bait, which is then eaten by pests, such as slugs, ants, and rodents.

If the same pesticide is used repeatedly, pests and diseases may develop resistance to it. Prevent this by employing a range of non-chemical measures and only using pesticides where absolutely necessary.

APPLICATION METHODS

Pesticides are available in a range of forms designed to maximize their efficacy and convenience.

Liquid pesticides are supplied in ready-to-use trigger-spray bottles – ideal for quickly treating small areas. They can also be bought as concentrates, which must be diluted before use. These are more economical than premixed sprays and easier to apply over larger areas, but their high strength means that they must be handled with great care and applied using the recommended spraying equipment. Liquid drenches, applied from a watering can, are used to treat soil-dwelling pests.

Pesticides may also be supplied as dusts, which can be more awkward to apply than sprays, and as pellets designed to be scattered around plants. Smoke fumigants can be useful to treat greenhouse pests.

Trigger spray bottles are a convenient way to deliver pesticides if you are treating a small patio or container plants.

SAFE AND EFFECTIVE USE

For the safety of people, pets, and wildlife, only use pesticides when absolutely necessary. Identify the problem and then choose a product that is suitable to treat the affected plant. Read and carefully follow the manufacturer's instructions printed on the label, paying attention to the correct timing of treatments and the intervals between them.

Pesticides may kill beneficial insects visiting plants at the same time that they act against pests. For this reason, never spray plants in flower to avoid killing pollinating insects. After spraying, keep people and pets away until the product is dry. Never eat, drink, or smoke when using pesticides and wash your hands thoroughly afterwards. Do not spray in windy weather to prevent spray drifting onto other plants, and keep pesticides away from water courses. Never mix different pesticides and clean all sprayers thoroughly after use to avoid accidental mixing or contamination. Store pesticides tightly closed in their original packaging.

TOP TIP BUY PESTICIDES IN SMALL QUANTITIES AND DILUTE THE RIGHT AMOUNT FOR A SINGLE USE TO AVOID HAVING TO DISPOSE OF THEM.

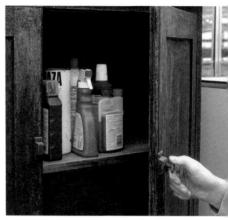

Store pesticides in a cupboard safely out of reach of children and pets.

TREATING EDIBLE CROPS

Many gardeners prefer not to use pesticides at all on the fruit, vegetables, and herbs that they intend to eat. If this is considered necessary, note that not all pesticides are approved for use on edible crops. Consult each product label carefully for instructions specifically relating to edible crops before use. Organic pesticides and plant invigorators (see pp.46–49) have a low toxicity and only remain on plants for a short time, so can be used on a wide range of edible plants, generally with a short interval between treatment and harvest. The greater toxicity and persistence of synthetic pesticides (see p.48) limits their use on edible plants. They must only be applied to the specific crops stated on the label, the quantity and frequency of treatments may be limited, and the interval between treatment and harvest will be longer.

Only spray small fruit trees where all branches can be thoroughly treated.

Plant invigorators (see p.48) can control insect pests on red cabbage.

CHECKING APPROVAL AND SAFE DISPOSAL

In most countries, it is illegal to use old pesticides that are no longer approved for amateur gardeners. In the UK, you can check to see if the old products you may have in your cupboards are still approved at the following website: **secure.pesticides.gov.uk/garden/prodsearch.asp**

To dispose of old pesticides, ensure they are safely sealed in their original packaging and hand them to staff at your local authority waste site. Never pour pesticides down the drain or sink.

PROTECTING WILDLIFE

- Use organic insecticides with short persistence to minimize the risks to beneficial insects.
- Never spray plants in flower.
- Don't use pesticides near ponds or water courses.
- Scatter slug pellets thinly to avoid poisoning other creatures.

CHOOSING PESTICIDES

If you do use pesticides to control a pest or disease, choosing the right product is essential to making the treatment safe and effective. Be sure to read product labels before purchase to ensure the chemical you use is recommended for the problem identified and the types of plants affected.

Pesticides contain either organic or synthetic active ingredients; these may have contact or systemic action (see p.44) and pose different levels of risk to beneficial insects. Relatively few ingredients are now approved for use by amateur gardeners, which makes it easier to select from what is available.

ORGANIC PESTICIDES

Organic pesticides contain active ingredients that are derived from natural sources, including plants and minerals. They have contact action, which means they need to touch pests directly to take effect, so all plant surfaces must be thoroughly sprayed. These chemicals can be used to control a range of invertebrate pests, including aphids, whitefly, and small caterpillars, but are also likely to kill beneficial insects that are accidentally sprayed. Short persistence means that their active ingredient is only present on plants for a brief period, usually from 24 hours to a few days. This means that repeated applications may be necessary to bring pests under control, but also makes organic pesticides much less likely to pollute the wider environment or harm beneficial invertebrates visiting plants after treatment.

Organic pesticides have relatively low toxicity to people and animals and are approved for use on many plants, including fruit and vegetables, where their short persistence allows them to be used close to harvest. However, always check the label to confirm that they are suitable for use on your crops.

Pyrethrum (*Tanacetum cinerariifolium*) flowers are the source of the organic pesticide pyrethrum.

ACTIVE INGREDIENT	APPLICATION	CONTROLS
Pyrethrum (flower extract); Pyrethrins (active molecules refined from extract)	Spray or dust	Many insects, including aphids, capsid bugs, caterpillars, thrips, and whitefly
Fatty acids (insecticidal soaps)	Spray	Small invertebrates, including aphids, leafhoppers, scale insects, thrips, and whitefly
Plant oils	Spray or winter wash	Small invertebrates, including aphids, red spider mites, scale insects, and whitefly
Garlic extract	Fumigant	Greenhouse insect pests, such as aphids and whitefly

SYNTHETIC PESTICIDES

The active ingredients of synthetic pesticides are made using chemical processes. They can be contact or systemic in action, but all have a long persistence on plants – several weeks, compared to a few days for organic pesticides. This makes a single application effective at controlling pests for a long period, but puts beneficial invertebrates at greater risk and means that a longer interval between treatment and harvest needs to be observed if they are used on edible plants.

Synthetic pyrethroids are contact pesticides based on the molecular structure of natural pyrethrum, with its low toxicity to mammals, but modified to persist for longer on plants. Acetamiprid is a common systemic insecticide that contains neonicotinoid compounds (see *right*), which are toxic to bees and other pollinating insects. Insecticides are sometimes combined with systemic fungicides in one product. You should only use combined products where treatment for both pests and fungal disease is necessary.

Handheld pump action sprayers are ideal for applying pesticides quickly and evenly.

TOP TIP PREPARE YOUR PESTICIDES FOR USE OUTDOORS OR IN A WELL-VENTILATED AREA, AND ONLY MIX THE AMOUNT YOU NEED FOR ONE APPLICATION. WEAR APPROPRIATE PROTECTIVE EQUIPMENT AND CLOTHING IF SPECIFIED.

ACTIVE INGREDIENT	APPLICATION	CONTROLS
Deltamethrin (synthetic pyrethroid)	Spray	Many invertebrates, including aphids, moth pests, flea beetles, sawfly larvae, and scale insects
Cypermethrin (synthetic pyrethroid)	Spray	Many invertebrates, including aphids, caterpillars, and beetles
Lambda-cyhalothrin (synthetic pyrethroid)	Spray	Many invertebrates, including aphids, beetles, caterpillars, pea moth and whitefly
Permethrin (synthetic pyrethroid)	Fumigant	Greenhouse insects, including aphids and whitefly
Acetamiprid (neonicotinoid)	Spray, compost drench	Many invertebrates, including aphids, caterpillars, and lily beetle. Drench treats vine weevil larvae
Myclobutanil (fungicide)	Spray	Black spot, powdery mildew, and rust
Tebuconazole (fungicide)	Spray	Black spot, box blight, powdery mildew, and rust
Trifloxystrobin (fungicide)	Soluble powder	Lawn fungal diseases
Triticonazole (fungicide)	Spray	Black spot, powdery mildew, and rust

NEONICOTINOID PESTICIDES

This contentious group of pesticides interferes with the nervous systems of insects, and their use has been implicated in the reduction of insect populations worldwide, including honey bees. Neonicotinoids remain present in plants for long periods after treatment, and are toxic to wildlife other than insects. Only acetamiprid is currently approved for amateur use in the UK, but gardeners should think carefully before using these chemicals in borders filled with flowering plants or on their fruit and vegetables.

CHEMICAL ALTERNATIVES

Commercial pesticides have their place in the control of pests and diseases. However, they are fairly costly and can have negative effects on wildlife and the environment. Other substances, made from more benign ingredients or that rely on physical (rather than chemical) properties to deter pests, provide alternatives for gardeners seeking more environmentally friendly options. Check product labels to avoid unwittingly using toxic ingredients. Researching home-made remedies can also be worthwhile, as many have been used by generations of gardeners for good reason.

Invigorators can be effective against mildew and a range of common insect pests.

PLANT INVIGORATORS

Products labelled plant invigorators are often marketed as "natural pesticides". They contain a combination of naturally derived ingredients that are less damaging to beneficial insects than conventional pesticides. They vary in composition but usually contain surfactants or oils that work by gumming up the wings and breathing tubes of insects. These are typically combined with nutrients that improve the overall health of plants. Invigorators are available as liquid concentrates or ready-to-use sprays. They need to be applied directly on to pests and on to all upper and lower leaf surfaces so that the whole plant is thoroughly wetted. Several applications might be needed before effects can be seen. These chemicals can be used on ornamental and edible plants.

SAFER SOLUTIONS

Scientific research continues to provide gardeners with safer alternatives for pest control. Slug pellets were once made mainly from metaldehyde, a chemical that is toxic to slugs and snails, but also to pets, birds, and wild mammals. It also washes through soil to pollute water courses.

Today, pellets made from ferric phosphate are available to buy. These are less toxic to wildlife, and also break down safely in the soil to produce iron and phosphate, which are nutrients that are taken up by plant roots.

Pellets containing ferric phosphate are approved for organic gardeners.

HOME-MADE REMEDIES

A quick search online will reveal numerous recipes for home-made pesticides that use plant extracts, kitchen ingredients, or other domestic products. Some gardeners use milk solution to combat powdery mildew, garlic drenches to repel slugs and snails, and soap spray to control sap-sucking insects. Others will sprinkle desiccants, such as diatomaceous earth, around plants to kill slugs and snails, or make pesticides by steeping leaves or flowers of plants, such as marigolds, in water, then dilute the resulting "tea" for use as a spray.

Some of these products are part of horticultural tradition while others derive from a holistic "biodynamic" approach to gardening. There is little formal research to support their effectiveness, but much anecdotal evidence to say they work. If you opt to try such approaches, be aware that recommended dilutions are often guesswork, as concentrations of home-made extracts will vary greatly.

TOP TIP SOME HOME-MADE REMEDIES MAY BE REGULATED: CHECK WITH YOUR LOCAL ENVIRONMENT HEALTH AUTHORITY BEFORE USE.

Many home-made recipes use some combination of garlic, washing-up liquid, and vegetable oil.

Always test home-made remedies on small areas of plants a week before applying them more widely.

Treatments for some caterpillars contain *Bacillus thuringiensis*: these are intended for professional use only.

SHOP SMART

Products to treat garden pests and diseases are subject to strict safety standards and licensed for sale by the relevant national authorities. You can buy with confidence from garden centres and hardware stores, but online retailers and auction sites are more difficult to regulate. You may find products for sale that are illegal in your country or that are intended for professional rather than amateur use.

Exercise caution when buying online. Packaging or instructions printed in a foreign language are a sign that the product may not be licensed for use in your country. Look for the list of active ingredients in the product description and check that they are permitted for use before buying. In the UK, you can find the relevant information at the following website: **secure.pesticides.gov.uk/garden/prodsearch.asp**

Many organic chemicals traditionally used as garden pesticides – sometimes for many decades – have had approval for amateur use withdrawn as regulations have tightened. Yellow sulphur, for example, was once widely used to treat powdery mildew and rose black spot, but can no longer be marketed as a fungicide in the EU and UK; however, it remains available for use as a soil acidifier and nutrient. Similarly, tar oils and tar acids were marketed as treatments for aphids and wine weevils within the last two decades, but have now been withdrawn from use. Keep an eye out for further changes to guidelines.

Give plants a healthy start and you will be much less likely to encounter problems. Finding out about the growing conditions they prefer is a great first step.

PLANT GROUPS

This chapter is divided into nine sections, each covering a widely grown group of plants. These sections contain advice on successful cultivation and tips on how to prevent problems that affect that group. Where a problem does arise, you can make a quick diagnosis by finding your plant's symptom in the table and checking the possible causes in the Directory of Pests, Diseases, and Disorders (*see pp.76–141*).

VEGETABLE CROPS

Plants grown for their flavour and nutrients unsurprisingly attract many pests and are susceptible to a range of diseases. Most gardeners prefer not to use chemical pesticides on food crops but rely on the principle that well-fed soil is the best route to healthy plants and generous yields. Many practical techniques can also be employed to prevent problems and to protect plants from pests.

Harvest crops as soon as they are ready to put them beyond the reach of pests.

Sow seeds at the time and spacing suggested on the seed packet.

GIVING A GOOD START

Pests and diseases can quickly overwhelm small seedlings, so it pays to promote vigorous early growth to minimize the time that your plants remain at this vulnerable stage. Begin by planting your seeds where the majority of veg plants will thrive – in sunny, sheltered borders or raised beds with free-draining soil, or in suitable containers. Wait until the soil has warmed up in spring before sowing seed outdoors, and closely follow the instructions on the seed packet. Many crops can be sown under cover to give them a head start at the seedling stage; harden them off gradually before planting outdoors.

PREVENTING PROBLEMS

Beyond encouraging your plants to "outgrow" the threats from pests and diseases, there are some simple measures you can take to minimize damage. Rotate vegetable crops (see p.33) because pests and diseases are far more likely to build up in your soil if you grow the same plants in the same spot year after year. For the same reason, don't reuse the compost from last year's containers. Improve your soil annually with a thick mulch of organic matter in autumn or spring, and if your plot has acidic soil, add lime in autumn. Not only will this make your soil more fertile, but it will reduce the incidence of clubroot on brassicas.

Keep your vegetable plot tidy by removing clutter, fallen leaves, spent plants, and weeds that could harbour pests and diseases. Don't risk planting any young plants, seed potatoes, or onion sets that show signs of pests or diseases that they could introduce to your plot.

Weeding your veg plot regularly not only makes it look tidier, but removes plants that may harbour pests and diseases and reduces competition for resources.

PROTECTING CROPS

Many pests can be kept at bay with physical barriers. Netting is ideal as it allows water, air, and light through, but it needs to be supported on a sturdy frame away from foliage and secured at the base so that pests can't get in. Cover brassicas with netting that has a maximum mesh size of 7mm (¼in) to keep out egg-laying butterflies. Place flat "collars" (see p.34) around bases of brassica stems to stop cabbage root flies laying their eggs in the soil.

To exclude small insects, such as flea beetles and carrot fly, you will need to use netting with a mesh size of 1.25mm (¹⁄₂₀in) or cover plants with cloches or horticultural fleece (which also insulates plants on cold nights).

Lines of eggshells or pine needles can deter slugs and snails but these pests are better lured away with beer traps. Sacrificial plants can also be used to attract slugs, snails, and aphids away from crops (see p.35).

A netted cage will protect brassicas from egg-laying butterflies as well as larger pests, such as pigeons.

CHOOSING VARIETIES

Many of the thousands of vegetable varieties on offer at garden centres and online have been bred with built-in resistance to diseases and some pests, making them exceptionally valuable to all gardeners, and especially those who choose not to use chemicals. It pays to study seed catalogues and select varieties that offer resistance to problems that you habitually encounter on your plot. Resistance doesn't guarantee complete protection for some problems, such as carrot fly and tomato blight, but it can be combined with other control methods to greatly improve your chances of success. Pathogens evolve, causing varieties to lose resistance to diseases like potato blight, so ensure you have up-to-date information. Look for bolt resistance in root and leafy crops and varieties bred to mature quickly in cooler regions.

Beetroot 'Boltardy' is a popular variety with excellent resistance to bolting from early sowings.

TYPES OF CROPS

Different pests and diseases affect different crop groups. Below are the groups referred to in the diagnostic table on pp.54–55.

BRASSICAS Brussels sprouts, cabbage, calabrese, cauliflower, sprouting broccoli, kale, radish, swede, turnip
FRUITING CROPS Aubergines, chillies, courgettes, cucumbers, peppers, squashes, tomatoes
LEAFY CROPS Chard, lettuce, spinach, rocket, mizuna
LEGUMES Borlotti, broad, French and runner beans, and peas
ROOT CROPS Beetroot, carrots, parsnips, potatoes
ALLIUMS Garlic, leeks, onions, shallots, spring onions

COMMON PROBLEMS

Find your vegetable crop's problem in the symptoms column, then scan across for possible causes, each of which is cross-referenced with detailed directory entries to help you make the right diagnosis.

SYMPTOMS	CROP GROUPS AFFECTED (SEE P.53)	POSSIBLE CAUSES
LEAVES		
Brown or grey-brown spots	Brassicas, Legumes, Root crops, Leafy crops, Fruiting crops, Alliums	Fungal leaf spot (see p.94) • Chocolate spot (see p.93) • Bacterial leaf spot (see p.92) • Bean halo blight (see p.92) • Weedkiller damage (see p.103)
Pale fuzzy growth from brown or yellow patches	Brassicas, Leafy crops, Legumes, Fruiting crops, Root crops, Alliums	Grey mould (see p.94) • Downy mildew (see p.93)
Insects visible	Brassicas, Legumes, Root crops, Leafy crops, Fruiting crops, Alliums	Aphids (see p.79) • Whitefly (see p.91) • Red spider mites (see p.86) • Thrips (see p.90) • Asparagus beetle (see p.80)
Larvae visible	Brassicas, Leafy crops, Fruiting crops	Caterpillars (see p.81)
Notches eaten from margins	Legumes	Pea and bean weevil (see p.84)
Raised blisters, warts or pustules	Brassicas, Legumes, Alliums, Fruiting crops, Leafy crops, Root crops	Rusts (see p.96) • Oedema (see p.102) • Smuts (see p.139)
Yellowing or other discolouration and weak growth	Brassicas, Legumes, Root crops, Leafy crops, Fruiting crops, Alliums	Potato blackleg (see p.138) • Carrot fly (see p.119) • Onion fly (see p.122) • Onion white rot (see p.125) • Boron deficiency (see p.100) • Nitrogen deficiency (see p.101) • Iron and manganese deficiency (see p.101) • Magnesium deficiency (see p.101) • Violet root rot (see p.127)
Brown or bleached, especially young growth	Brassicas, Legumes, Alliums, Fruiting crops, Leafy crops, Root crops	Frost damage (see p.100) • Drought (see p.100) • Wind damage (see p.103) • Scorch (see p.103)
Tunnels or pale patches	Root crops, Alliums, Leafy crops	Leaf miners (see p.83) • Leek moth (see p.83)
Holes or entire leaves eaten	Brassicas, Legumes, Root crops, Leafy crops, Fruiting crops, Alliums	Flea beetles (see p.82) • Caterpillars (see p.81) • Slugs and snails (see p.88) • Rabbits (see p.85) • Deer (see p.132) • Pigeons (see p.85) • Bean seed fly (see p.80)
Yellow mottling and stunted growth	Brassicas, Legumes, Root crops, Leafy crops, Fruiting crops, Alliums	Viruses (see p.99)
White, dusty coating	Brassicas, Fruiting crops, Leafy crops, Root crops, Legumes	Powdery mildew (see p.95)
Small leaves on rapidly extending flowering stem	Brassicas, Root crops, Leafy crops, Root crops, Alliums	Bolting (see p.108)
Wilting and sometimes yellowing	Brassicas, Legumes, Root crops, Leafy crops, Fruiting crops, Alliums	Verticillium wilts (see p.139) • Foot and root rots (see p.124) • Root aphids (see p.122) • Clubroot (see p.124) • Cabbage root fly (see p.119) • Drought (see p.100)
Spreading brown patches	Root crops, Fruiting crops	Potato blight (see p.95) • Tomato blight (see p.97)
Discoloured and deformed foliage	Brassicas, Legumes, Root crops, Leafy crops, Fruiting crops, Alliums	Weedkiller damage (see p.103)
FLOWERS AND BUDS		
Insects visible	Legumes, Fruiting crops	Aphids (see p.105)
Holes in petals	Legumes, Fruiting crops	Slugs and snails (see p.106)
Brown patches and fuzzy growth	Legumes, Fruiting crops	Grey mould (see p.107)
Buds brown and fail to open, petals turn brown	Legumes, Fruiting crops	Frost damage (see p.108) • Drought (see p.100)

SYMPTOMS	CROP TYPES AFFECTED (SEE P.53)	POSSIBLE CAUSES
FRUITS AND PODS		
Larvae tunnelling inside	Legumes, Fruiting crops	Pea moth (see p.112) • Tomato moth (see p.113)
Brown patches and fuzzy growth	Legumes, Fruiting crops	Grey mould (see p.115) • Tomato blight (see p.115)
Fruit skin split	Fruiting crops	Fruit split (see p.117)
Black or dark brown patches at base of fruit	Fruiting crops	Blossom end rot (see p.116)
Green or pale marks on ripening tomatoes	Fruiting crops	Tomato greenback (see p.117)
Holes or partially eaten	Legumes, Fruiting crops	Mice (see p.112) • Slugs and snails (see p.88)
Unpleasant-tasting cucumbers	Fruiting crops	Bitter cucumbers (see p.116)
Few fruits or pods forming	Legumes, Fruiting crops	Fruit fails to set (see p.117)
ROOTS		
Tunnels in root and bulb crops, larvae may be visible	Root crops, Alliums, Brassicas	Cabbage root fly (see p.119) • Carrot fly (see p.119) • Onion fly (see p.122) • Wireworms (see p.123) • Slugs (see p.122)
Larvae feeding on roots or seeds in soil	Brassicas, Alliums, Root crops, Legumes	Bean seed fly (see p.119) • Cabbage root fly (see p.119) • Onion fly (see p.122) • Cutworms (see p.120) • Chafer grubs (see p.120)
Pale insects feeding on roots	Legumes, Leafy crops, Root crops	Root aphids (see p.122)
Roots swollen and deformed	Brassicas	Clubroot (see p.124)
Brown patches on root crops	Root crops	Parsnip canker (see p.125) • Potato blight (see p.126) • Potato common scab (see p.126) • Potato gangrene (see p.126) • Powdery potato scab (see p.127) • Potato spraing (see p.127)
Silvery-grey marks on potato tubers	Root crops	Potato silver scurf (see p.127)
Dark purple strands cover roots and tubers	Root crops, Fruiting crops	Violet root rot (see p.127)
Roots rotten, white fungal growth at base of bulb	Alliums	Onion white rot (see p.125)
Roots brown and rotten	Legumes, Fruiting crops, Alliums, Leafy crops	Foot and root rots (see p.124)
Centre of root crops hollow or rotten	Root crops, Brassicas	Potato hollow heart (see p.129) • Boron deficiency (see p.128)
Roots split or malformed	Root crops, Brassicas	Boron deficiency (see p.128) • Root fanging (see p.129) • Root split (see p.129)
STEMS		
Stems of seedlings and young plants collapse	Root crops, Alliums, Legumes, Brassicas, Leafy crops, Fruiting crops	Damping off (see p.135) • Cutworms (see p.120) • Slugs (see p.122)
Insects visible	Root crops, Legumes, Brassicas, Leafy crops, Fruiting crops	Aphids (see p.131) • Asparagus beetle (see p.131)
Stems bruised, bent, or broken	Legumes, Fruiting crops, Leafy crops, Brassicas	Wind damage (see p.141)
Black or brown marks on stems, sometimes with fungal growth	Root crops, Alliums, Legumes, Brassicas, Leafy crops, Fruiting crops	Potato black leg (see p.138) • Potato blight (see p.95) • Smuts (see p.139) • Grey mould (see p.137) • Foot and root rots (see p.136)
Stems grazed	Root crops, Alliums, Legumes, Brassicas, Leafy crops, Fruiting crops	Deer (see p.132) • Rabbits (see p.132)

FRUIT CROPS

Homegrown fruit, picked perfectly ripe, tastes far superior to anything available in the shops and is surprisingly easy to fit into even the smallest garden. Careful cultivation, regular pruning, and good garden hygiene all help to ensure heavy crops and prevent diseases spreading. Although many pests have a taste for sweet fruit, there are plenty of effective ways to protect crops without the use of chemicals.

Redcurrants are prone to damage by the gooseberry sawfly.

Pick fruits as soon as they ripen, before birds get a chance to feast on them.

CULTIVATION

Fruit crops thrive in a sunny position, sheltered from cold winds and well away from frost pockets. They need fertile, well-drained soil that has been improved with organic matter before planting. Trees, bushes, and cane fruit are all best bought as young plants and planted between early winter and early spring, while strawberries should be planted in late summer for a crop the following year. Grow trees and bushes as free standing plants, trained against a wall or fence to save space, or even in containers. Fruit will ripen faster when grown against a sunny, south-facing wall. Check fruit regularly and harvest when ripe to reduce the opportunities for pests or diseases to strike.

PREVENTING PROBLEMS

Buy plants from a specialist fruit nursery or reputable retailer who can supply stock that has been certified disease-free under the UK Plant Health Propagation Scheme (PHPS). Fruiting plants need a steady supply of water and nutrients through the growing season to enable them to bear a good crop. Mulch and feed plants with a balanced fertilizer in spring, and water during dry spring and summer weather to allow fruit to set and swell. Remove weeds regularly to prevent them harbouring pests and diseases and competing with plants. Prune each crop carefully and at the correct time, to avoid removing fruiting spurs. Use sharp secateurs to make clean cuts that will heal quickly, and disinfect tools between plants to avoid spreading disease. Rake up fallen leaves and fruits that show signs of pests and diseases promptly, and remove rotten fruits from branches to stop pathogens spreading or overwintering.

When pruning blackcurrant bushes, cut out dead or damaged wood close to ground level to improve airflow, let in more light, and promote new growth.

PROTECTING CROPS

Birds are the most troublesome fruit pests. The only sure way to stop them from pecking tree fruit or feasting on your berries and currants is to cover plants with netting. To do this, you can grow fruit bushes inside a permanent fruit cage, cover plants with netting well before the fruit starts to ripen, or protect selected branches or clusters of ripening fruit on trees with fleece jackets. Always make sure that netting is supported above the crops to stop birds pecking through it; also, secure the netting at its base to ensure that birds don't get trapped within. Bird scarers are far less reliable than netting and need to be moved frequently to have any effect.

A variety of traps and treatments, including pheromone traps, grease bands, and winter washes, will reduce the damage caused by insect pests and minimize the need for chemical controls. To prevent frost damage, cover trees and bushes with fleece overnight when in blossom.

Grease bands prevent wingless winter moths from climbing up fruit tree trunks and laying their eggs on leaves.

Cover a strawberry patch with netting to stop birds taking fruit.

Push rubber tops on to canes supporting netting to prevent injury.

CHOOSING VARIETIES

Fruit bushes and trees will be with you for years or even decades, so it is worth taking time to select varieties that align with your taste, local growing conditions, and storage requirements. Late-flowering varieties suit colder areas, while others have been bred to tolerate high rainfall. Many varieties have been developed for disease resistance – invaluable if you opt to grow your fruit without the use of pesticides.

The rootstock onto which tree varieties are grafted may provide disease resistance and will also dictate size. Dwarfing rootstocks, such as M9 for apples and Pixy for plums, keep trees compact for small gardens. Check whether your bushes or trees need to be planted with a pollination partner (a variety of the same species) to set fruit.

If growing apples in a cooler region, choose a good later-flowering variety such as 'Laxton's Superb'.

TYPES OF CROPS

Different pests and diseases affect different fruit groups. Below are the groups referred to in the diagnostic table on pp.58–59.

STONE FRUIT TREES Cherry, plum, greengage, damson, peach, nectarine, apricot
POME FRUIT TREES Apple, pear, quince
FRUIT BUSHES Blackcurrant, redcurrant, white currant, gooseberry, blueberry
CANE FRUITS Raspberry, blackberry, loganberry and other hybrids
STRAWBERRIES
GRAPE VINES
FIGS

COMMON PROBLEMS

Find your fruit crop's problem in the symptoms column, then scan across for possible causes, each of which is cross-referenced with detailed directory entries to help you make the right diagnosis.

SYMPTOMS	CROP GROUPS AFFECTED (SEE P.57)	POSSIBLE CAUSES
LEAVES		
Brown or grey-brown spots	All fruit crops	Apple and pear scabs (see p.114) • Bacterial leaf spot (see p.92) • Fungal leaf spot (see p.94) • Weedkiller damage (see p.103)
Pale fungal growth from brown or yellow patches	Strawberries, Grape vines	Grey mould (see p.94) • Downy mildew (see p.93)
White, dusty coating	Stone fruit trees, Pome fruit trees, Fruit bushes, Cane fruits, Strawberries, Grape vines	American gooseberry mildew (see p.114) • Downy mildew (see p.93) • Powdery mildew (see p.95)
Insects visible	All fruit crops	Aphids (see p.79) • Apple suckers (see p.105) • Leafhoppers (see p.83) • Red spider mites (see p.86) • Scale insects (see p.87) • Whitefly (see p.91)
Larvae visible	All fruit crops	Caterpillars (see p.81) • Sawflies (see p.87) • Winter moth (see p.91)
Notches eaten from margins	Strawberries	Vine weevil (see p.91)
Tunnels or pale patches	Stone fruit trees, Pome fruit trees	Leaf miner (see p.83)
Colourful raised blisters or pustules	Stone fruit trees, Pome fruit trees, Fruit bushes, Cane fruits, Strawberries	Aphids (see p.79) • Peach leaf curl (see p.94) • Pear leaf blister mite (see p.85) • Rusts (see p.96)
Yellowing or other discolouration and weak growth	All fruit crops	Boron deficiency (see p.100) • Nitrogen deficiency (see p.101) • Iron and manganese deficiency (see p.101) • Magnesium deficiency (see p.101) • Phosphorus deficiency (see p.102) • Potassium deficiency (see p.101)
Leaves turn brown especially at shoot tips	All fruit crops	Blossom wilt (see p.100) • Drought (see p.100) • Fireblight (see p.94) • Frost damage (see p.100) • Scorch (see p.103) • Wind damage (see p.103)
Holes or entire leaves eaten	All fruit crops	Apple capsid (see p.111) • Caterpillars (see p.81) • Deer (see p.132) • Rabbits (see p.85) • Sawflies (see p.87) • Shothole (see p.97) • Slugs and snails (see p.88) • Winter moth (see p.91)
Wilting, sometimes yellowing and leaf fall	All fruit crops	Drought (see p.100) • Foot and root rots (see p.124) • Honey fungus (see p.137) • Phytophthora root rot (see p.125) • Wilting (see p.91)
Silvery sheen on upper surface	Stone fruit trees	Silver leaf (see p.97)
Deformed and sometimes discoloured	All fruit crops	Aphids (see p.79) • American gooseberry mildew (see p.114) • Eelworms (see p.82) • Viruses (see p.99) • Weedkiller damage (see p.103)
Black covering on upper surfaces	Stone fruit trees, Pome fruit trees, Fruit bushes	Aphids (see p.79) • Scale insects (see p.87) • Whitefly (see p.91)
FLOWERS AND BUDS		
Buds or blossom missing or containing holes	Stone fruit trees, Pome fruit trees, Fruit bushes, Cane fruits, Strawberries	Birds (see p.105) • Slugs and snails (see p.106)
Flowers deformed	Stone fruit trees, Pome fruit trees, Fruit bushes, Cane fruits, Strawberries	Aphids (see p.105) • Apple sucker (see p.105) • Viruses (see p.107)
Brown patches and fuzzy fungal growth	Stone fruit trees, Pome fruit trees, Fruit bushes, Cane fruits, Strawberries	Grey mould (see p.107)
Buds brown and fail to open, petals turn brown	All fruit crops	Blossom wilt (see p.107) • Drought damage (see p.108) • Fireblight (see p.94) • Frost damage (see p.108)

SYMPTOMS	CROP TYPES AFFECTED (SEE P.57)	POSSIBLE CAUSES
FRUITS		
Larvae feeding and/or tunnelling inside	Stone fruit trees, Pome fruit trees, Cane fruit	Apple sawfly (see p.111) • Codling moth (see p.111) • Pear midge (see p.113) • Plum moth (see p.113) • Raspberry beetle (see p.113)
Raised blemishes on skin	Pome fruit trees	Apple capsid (see p.111) • Apple sawfly (see p.111)
Holes eaten in fruit flesh	All fruit crops	Birds (see p.111) • Mice (see p.112) • Slugs and snails (see p.88) • Strawberry seed beetle (see p.113) • Wasps (see p.113)
Fruit drops early or remains small	All fruit crops	Aphids (see p.79) • Drought (see p.108) • Pear midge (see p.113)
White or brownish powdery coating	Stone fruit trees, Fruit bushes, Cane fruits, Strawberries, Grape vines, Figs	American gooseberry mildew (see p.114) • Powdery mildew (see p.115)
Dark brown spots on skin	Pome fruit trees	Apple and pear scab (see p.114) • Apple bitter pit (see p.116)
Fruits turn brown and rot	All fruit crops	Brown rot (see p.114) • Grey mould (see p.115)
Fruit skin cracked or split	All fruit crops	Apple and pear scab (see p.114) • American gooseberry mildew (see p.114) • Fruit splitting (see p.117) • Irregular watering (see p.140) • Powdery mildew (see p.115)
Brown flecks in flesh	Pome fruit trees	Apple bitter pit (see p.116)
Plums pale yellow and distorted	Stone fruit trees	Pocket plum (see p.115)
Few fruits form	All fruit crops	Apple sucker (see p.105) • Drought (see p.108) • Frost damage (see p.108) • Fruit fails to set (see p.117)
ROOTS		
Roots brown and soft	All fruit crops	Foot and root rots (see p.124) • Phytophthora root rot (see p.125) • Waterlogging (see p.129)
Larvae or tiny worms feeding	Strawberries, Fruit bushes	Eelworms (see p.120) • Vine weevils (see p.123)
Rounded swellings on roots	Stone fruit trees, Pome fruit trees, Fruit bushes, Cane fruits	Crown gall (see p.124)
STEMS		
Stems grazed	All fruit crops	Deer (see p.132) • Rabbits (see p.132)
Small insects visible	All fruit crops	Aphids (see p.131) • Woolly aphids (see p.133)
Clusters of domed brown shells	Stone fruit trees, Pome fruit trees, Fruit bushes, Cane fruits, Grape vines, Figs	Scale insects (see p.133)
Leaf buds swollen and fail to open	Fruit bushes	Blackcurrant big bud mite (see p.131)
Stems turn brown and die back	All fruit crops	Dieback (see p.136) • Drought (see p.140) • Fireblight (see p.94) • Foot and root rot (see p.124) • Frost damage (see p.140) • Honey fungus (see p.137) • Poor pruning (see p.141) • Spur blight (see p.139) • Verticillium wilts (see p.139) • Waterlogging (see p.141) • Weedkiller damage (see p.141) • Wind damage (see p.141)
Splits in bark	Stone fruit trees, Pome fruit trees, Fruit bushes, Cane fruits, Grape vines	Crown gall (see p.124) • Irregular watering (see p.140)
Bark swollen and sometimes roughened	Stone fruit trees, Pome fruit trees, Fruit bushes, Cane fruits, Grape vines	Crown gall (see p.124) • Fungal canker (see p.137)
Small orange or buff pustules on dead wood	Stone fruit trees, Pome fruit trees, Fruit bushes, Cane fruits, Grape vines, Figs	Coral spot (see p.135) • Spur blight (see p.139)
Sunken, oozing areas of bark	Stone fruit trees	Bacterial canker (see p.134)

TREES

Trees are spectacular structural plants, creating striking forms clothed with beautiful leaves, delicate blossom, and vibrant berries. They add diversity to any garden, providing valuable shade and shelter from the wind. Trees will often shrug off pests and diseases, but watch out for early leaf fall, dieback on branches, or weeping wounds, which may indicate poor growing conditions or more serious disease.

The white trunks of *Betula* trees provide dramatic contrast in the garden.

Use a stick when planting to check that the top of the rootball is at soil level.

CULTIVATION

Select trees to suit your climate, growing conditions, and the space available, because they will be difficult to move and their natural forms may be spoiled if pruning is needed to control their size. Plant when the weather is cool and the soil is moist, from autumn to early spring. Dig a hole about three times as wide as the rootball and to the same depth as the soil mark on the trunk. Place the tree in the centre and firm soil back around its roots, so that there are no air pockets, and water thoroughly. Trees can also be planted in large containers, provided that they are well-drained and filled with soil-based compost.

PREVENTING PROBLEMS

Trees are relatively undemanding once they are established. Water new trees generously every two weeks during their first growing season, particularly during hot, dry weather, and add a thick layer of organic mulch over the soil at their base each spring. Always secure tall specimens or those on windy sites to a sturdy stake with a special tree tie to keep them stable, and add tree guards around trunks where pests like deer may gnaw the bark.

Pests and diseases can often be found on trees, but are usually of little concern and easily tolerated on such large plants. Watch for signs of more serious problems though, like dieback from the shoot tips or roughened cankers on branches, because the spread of disease can often be prevented if affected areas are pruned out quickly. Yellowing or falling foliage may be a sign of physiological stress or nutrient deficiency and should not be ignored.

Pruning out dead, diseased, or damaged wood can help to keep trees healthy, but take care to use sterilized tools and make clean cuts just above a bud or at the branch collar, away from the trunk, to encourage rapid healing and prevent the cuts from providing entry points for infection. Most deciduous trees are best pruned from late autumn to late winter, but leave ornamental cherries until summer to avoid silver leaf disease (see p.97).

Prune trees to remove dead or diseased branches, and allow in more air and light to promote healthy growth.

COMMON PROBLEMS

Find your tree's problem in the symptoms column, then scan across for possible causes, each of which is cross-referenced with detailed directory entries to help you make the right diagnosis.

SYMPTOMS	POSSIBLE CAUSES
LEAVES	
Dark brown or black spots or patches	Ash dieback (see p.134) • Bacterial leaf spot (see p.92) • Fungal leaf spot (see p.94) • Apple and pear scab (see p.114) • Pestalotiopsis (see p.95)
Sooty mould on upper surfaces	Aphids (see p.79) • Scale insects (see p.87)
Raised lumps on surfaces	Gall mites (see p.82)
Pale tunnels within leaf	Leaf miners (see p.83)
Circular holes and brown spots	Shothole (see p.97)
Browning from tips or edges	Frost damage (see p.100) • Scorch (see p.103) • Drought (see p.100) • Ramorum dieback (see p.96)
Raised orange or brown pustules	Rusts (see p.96)
Browning and dying back from shoot tips	Bacterial canker (see p.134) • Blossom wilt (see p.92) • Fireblight (see p.94) • Waterlogging (see p.141)
White powder on surface	Powdery mildew (see p.95)
Insects visible	Aphids (see p.79) • Scale insects (see p.87)
Larvae or silky webbing visible	Tortrix moth (see p.90) • Caterpillars (see p.81)
Yellowing and/or dropping foliage	Iron and manganese deficiency (see p.101) • Magnesium deficiency (see p.101) Drought (see p.100) • Waterlogging (see p.141) • Phytophthora root rot (see p.125)
Deformed at shoot tips	Bay sucker (see p.80) • Aphids (see p.79) • Weedkiller damage (see p.103)
Silver sheen on surface	Silver leaf (see p.97)
FLOWERS AND BUDS	
Brown patches	Frost damage (see p.108) • Grey mould (see p.107)
Wilt and wither	Blossom wilt (see p.107) • Fireblight (see p.94)
FRUITS	
Brown patches with fungal growth	Grey mould (see p.107) • Brown rot (see p.114)
Berries eaten	Birds (see p.111)
ROOTS	
Fine roots rotted and large roots blackened	Phytophthora root rot (see p.125) • Waterlogging (see p.129)
Rounded swellings on roots or base of tree	Crown gall (see p.124)
STEMS	
Orange pustules on dead wood	Coral spot (see p.135)
Dieback	Ash dieback • (see p.134) • Pestalotiopsis (see p.95) • Spur blight (see p.139) • Fireblight (see p.94) • Honey fungus (see p.137) • Drought (see p.140) • Phytophthora root rot (see p.125)
Toadstools at base or growing from stem	Honey fungus (see p.137) • Bracket fungus (see p.134)
Abnormally dense clusters of twigs	Witches' brooms (see p.139)
Bark gnawed	Deer (see p.132) • Squirrels (see p.133) • Rabbits (see p.132)
Bark splits or roughens and may ooze liquid	Bacterial canker (see p.134) • Fungal canker (see p.136) • Ramorum dieback (see p.96) • Ash dieback (see p.134)

SHRUBS

Shrubs can be deciduous or evergreen, giant or diminutive in size. Valued for their varied foliage and vibrant flowers and berries, they are generally robust, but may be prone to nutrient deficiencies, frost damage, insect pests, and root infections. Watch out for yellowing, browning, or damaged leaves and dieback of shoots or branches, which are all common signs of problems.

Shrubs are woody-stemmed plants that branch out close to ground level.

CULTIVATION

Choosing shrubs that suit your growing conditions – including aspect, climate, and soil pH – is the first step to keeping them healthy. Allow space for each plant to grow to maturity and improve the soil with plenty of well-rotted compost or manure before planting.

Plant shrubs at the same depth as they were growing in their pot and water regularly during the first growing season. After that, an annual mulch of garden compost in late winter, plus a feed with fertilizer (if required), should keep them growing happily. Shrubs in large containers need regular watering and feeding during the growing season.

Firm the soil around the roots to prevent your shrub rocking in the wind.

Blackflies can infest new leaves, but are usually controlled by natural predators.

PREVENTING PROBLEMS

Opt for disease-resistant cultivars where they are available. If you know that a particular pest or disease prevails in your garden, select plants that it will not affect; for example, grow a hedge of yew or berberis rather than box if box blight is a known problem. Where large pests graze on shrubs, erect a fence to exclude them from the garden or enclose newly planted shrubs for protection.

Be vigilant and check regularly for signs of trouble. Watch soft shoot tips for insect pests, holes, and deformed growth, and leaf surfaces for sooty mould. Discoloured leaves and dieback could be signs of disease and, if no immediate cause can be found, check

Sterilize secateurs with household disinfectant after use on diseased plants.

conditions at the roots. Rake up fallen leaves from beneath shrubs showing signs of pests or diseases, to prevent spores or eggs overwintering to return the following year.

Routine pruning helps to increase air flow among branches and can encourage attractive flowers and new growth. Pruning out dead or damaged stems also prevents them becoming an easy access point for disease, while the removal of diseased shoots can help limit the spread of infection. Vigorous reverted green shoots (see p.15) on variegated plants should be cut out promptly to prevent them becoming dominant. Use clean, sharp tools to make each pruning cut so that wounds heal quickly and disease is not spread between plants.

COMMON PROBLEMS

Find your shrub's problem in the symptoms column, then scan across for possible causes.
Each of these is cross-referenced with a detailed description to help you make the right diagnosis.

SYMPTOMS	POSSIBLE CAUSES
LEAVES	
Holes, sections, or whole leaves missing	Caterpillars (see p.81) • Sawflies (see p.87) • Viburnum beetles (see p.90) • Vine weevils (see p.91)
Discoloured areas	Leaf miners (see p.83) • Downy mildew (see p.93)
Puckered or deformed, especially at shoot tips	Aphids (see p.79) • Box/bay suckers (see p.80) • Capsid bugs (see p.80) • Gall mites (see p.82) • Viruses (see p.99) • Weedkiller damage (see p.103)
Sooty mould on upper surfaces	Aphids (see p.79) • Scale insects (see p.87) • Whiteflies (see p.91)
Insects visible	Aphids (see p.79) • Rosemary beetles (see p.86) • Thrips (see p.90) • Viburnum beetles (see p.90) • Vine weevils (see p.91) • Whiteflies (see p.91)
Larvae visible	Caterpillars (see p.81) • Sawflies (see p.87) • Winter moth (see p.91)
Silk webbing binding leaves together	Tortrix moth (see p.90) • Winter moth (see p.91)
Grey or brown spots	Bacterial leaf spot (see p.92) • Fungal leaf spot (see p.94)
Brown patches or edges	Drought (see p.100) • Scorch (see p.103) • Ramorum dieback (see p.96)
Leaves turning brown from shoot tips	Box blight (see p.93) • Fireblight (see p.94) • Pestalotiopsis (see p.95) • Frost damage (see p.100)
White powder on surface	Powdery mildew (see p.95)
Raised orange/brown pustules	Rusts (see p.96)
Yellowing leaves and/or leaf drop	Drought (see p.100) • Weedkiller damage (see p.103) • Nitrogen deficiency (see p.101) • Iron and manganese deficiency (see p.101) • Magnesium deficiency (see p.101) • Potassium deficiency (see p.102) • Waterlogging (see p.129) • Phytophthora root rot (see p.125) • Honey fungus (see p.137)
FLOWERS AND BUDS	
Wilt and wither	Fireblight (see p.94) • Honey fungus (see p.137)
Brown patches	Frost damage (see p.108) • Grey mould (see p.107) • Rose balling (see p.109)
Buds distort and fail to develop	Capsid bugs (see p.105)
FRUITS	
Soft brown patches with grey fuzzy growth	Grey mould (see p.115)
Berries eaten	Birds (see p.111)
ROOTS	
Fine roots rotted and large roots blackened	Phytophthora root rot (see p.125) • Waterlogging (see p.129)
Rounded swellings appear	Crown gall (see p.124)
STEMS	
Die back	Box blight (see p.93) • Fireblight (see p.94) • Pestalotiopsis (see p.95) • Frost damage (see p.140) • Phytopthora root rot (see p.125) • Weedkiller damage (see p.141) • Honey fungus (see p.137)
Spreading lesions	Ramorum dieback (see p.96)
Small, smooth, brown lumps	Scale insects (see p.133)
Toadstools around base	Honey fungus (see p.137)

CLIMBERS

Perfect for clothing walls and fences with their ornamental foliage and showy flowers, climbers tend to be tough, vigorous plants, but most need training on to additional supports to help them look their best. Climbers tend to be resilient but are often grown against walls and fences, where they may be at risk of drought and nutrient deficiencies. Their dense growth also provides ideal cover for pests.

Clematis HAPPY BIRTHDAY is prone to clematis wilt.

Hydrangeas are a good choice for a north- or east-facing wall.

CULTIVATION

Before planting, check the aspect of the wall or fence. Shady, north-facing walls suit some climbers, but will prevent sun-lovers, such as wisteria, from flowering. Choose tough climbers like ivy for positions exposed to strong winds. Improve the soil with well-rotted compost and dig a hole twice as wide as the rootball, preferably 30cm (12in) from the wall or fence so rain can the reach the roots. For most plants, position the rootball so its top is level with the soil surface, but sink clematis 10cm (4in) deeper into the soil: this will help them recover if they are later struck by clematis wilt (see p.93).

PREVENTING PROBLEMS

Some climbers are self-clinging and will attach themselves to their supporting structure, but many others are twining or scrambling plants that will need wires, mesh, or sturdy trellis around which to anchor their stems as they grow. Find out what type of climber you are buying so that you can choose and erect suitable supports.

Gently tie in the stems of new climbers to the supports to get them started. Continue to train and tie in new stems as the plants grow to prevent the wind from whipping and breaking them. Check and loosen the ties as the plants mature to stop them cutting into stems and allowing in disease.

Most significant problems with climbers are caused by dry or impoverished soil, where the plants have been established in small gaps between paving and walls, which rarely receive any rain. This can lead to yellowing foliage caused by drought and nutrient deficiency. Such weakened plants will readily succumb to powdery mildew and aphids, and are unlikely to flower well. Insect pests can also flourish in the warm sheltered conditions against sunny walls.

An organic mulch in spring will help to improve soil, while regular watering and feeding through spring and summer will promote healthy growth and flowering. Check the pruning requirements for your climbers and make cuts using clean, sharp tools so that wounds heal quickly.

Morning glory stems will twine around plant supports.

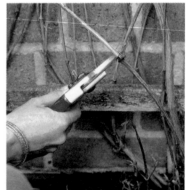

Clematis falls into one of three groups depending on its pruning needs.

COMMON PROBLEMS

Find your climber's problem in the symptoms column, then scan across for possible causes, each of which is cross-referenced to detailed directory entries to help you make the right diagnosis.

SYMPTOMS	POSSIBLE CAUSES
LEAVES	
Wilting	Clematis wilt (see p.93) • Drought (see p.100) • Verticillium wilts (see p.139)
Insects visible	Aphids (see p.79) • Whitefly (see p.91) • Thrips (see p.90) • Red spider mites (see p.86) Scale insects (see p.87)
Webbing on foliage	Red spider mites (see p.86) • Caterpillars (see p.81)
Larvae visible	Caterpillars (see p.81)
Powdery white coating	Powdery mildew (see p.95)
Black sooty mould on upper surface	Aphids (see p.79) • Whitefly (see p.91) • Scale insects (see p.87)
Brown edges or tips	Drought (see p.100) • Frost damage (see p.100) • Scorch (see p.103)
Yellowing and/or leaf drop	Iron and manganese deficiency (see p.101) • Magnesium deficiency (see p.101) • Nitrogen deficiency (see p.101) • Drought (see p.100) • Viruses (see p.99) • Red spider mites (see p.86) • Phytophthora root rot (see p.125)
Distorted, often at shoot tips	Aphids (see p.79) • Capsid bugs (see p.80) • Viruses (see p.99)
Holes or notches in leaves	Caterpillars (see p.81) • Vine weevils (see p.91) • Slugs and snails (see p.88) • Flea beetles (see p.82) • Capsid bugs (see p.80)
Brown or grey spots	Fungal leaf spot (see p.94)
Pale tunnels or blotches	Leaf miners (see p.83)
Pale mottling on upper surface	Thrips (see p.90) • Red spider mites (see p.86)
FLOWERS AND BUDS	
Turn brown with fuzzy fungal growth	Grey mould (see p.107)
Sudden wilting	Clematis wilt (see p.93)
Brown edges or patches	Frost (see p.108) • Wind damage (see p.109)
Insects visible	Aphids (see p.105) • Thrips (see p.90) • Earwigs (see p.106)
FRUITS	
Berries eaten	Birds (see p.111)
ROOTS	
Fine roots rotted and large roots blackened	Phytophthora root rot (see p.125) • Waterlogging (see p.129)
STEMS	
Wilting that doesn't recover with watering	Clematis wilt (see p.93) • Verticillium wilts (see p.139)
Clusters of small, raised brown lumps	Scale insects (see p.133)
Dieback	Honey fungus (see p.137) • Phytophthora root rot (see p.125) • Waterlogging (see p.129) • Frost (see p.140)
Bent or broken	Wind damage (see p.141)
Toadstools around base	Honey fungus (see p.137)

PERENNIALS

Perennials are non-woody plants. Some are evergreen but most die back in autumn to regrow in spring. If they suit your garden's growing conditions, they will put on healthy growth and vibrant displays of flowers. Most problems are fleeting, as growth is renewed each year, but pests can damage the soft spring shoots, while fungal diseases affect plants weakened by waterlogged soil or summer droughts.

Hellebores bear elegant winter blooms but are prone to leaf spot.

CULTIVATION

Always choose perennials that are adapted to flourish in your soil and climate, and plant them in spring or autumn, when the soil is moist. Before planting, weed thoroughly and improve soil with plenty of well-rotted compost. Space plants to allow for growth so that air can circulate between them and reduce the likelihood of fungal diseases. Dig a hole just wide and deep enough for the rootball, and position the base of the shoots or stems just above soil level, to prevent them rotting. Gently firm soil around the roots and water thoroughly. Perennials also thrive in containers with good drainage.

Use perennials to create spectacular and long-lasting container displays.

Deter slugs from hosta foliage with a copper band around the pot rim.

PREVENTING PROBLEMS

Most perennials are tough and don't suffer serious problems, but they can be vulnerable to pests as their tender new shoots emerge in spring. Take measures to control slugs and snails, as well as grazing rabbits or deer, before damage occurs and pick off or squash any insect pests immediately. Feed the soil with a thick annual mulch of compost each spring to produce steady, sturdy growth, because fertilizers can promote soft shoots that are a magnet for pests.

Plants stressed by difficult growing conditions are more likely to succumb to diseases like powdery mildew, so water regularly after planting and during any long dry spells of weather. Reduce fungal

Dividing perennials in spring or autumn rejuvenates congested old plants.

diseases associated with wet conditions, including downy mildew and foot and root rots, by choosing suitable perennials for heavy or persistently damp soil. Varieties bred with disease resistance are also often available.

Add supports in spring for tall perennials to grow through to prevent wind breaking their stems. Remove diseased growth or fallen leaves to prevent problems spreading and always cut back faded foliage and flower heads before spring growth begins in order to remove any overwintered pests and diseases. Lifting and dividing clumps of perennials every 3–5 years helps boost health by allowing weeds to be removed and healthy young sections to be selected for replanting.

COMMON PROBLEMS

Find your perennial's problem in the symptoms column, then scan across for possible causes, each of which is cross-referenced to detailed directory entries to help you make the right diagnosis.

SYMPTOMS	POSSIBLE CAUSES
LEAVES	
Insects visible	Aphids (see p.79) • Leafhoppers (see p.83) • Red spider mites (see p.86)
Larvae visible	Caterpillars (see p.81) • Swift moth larvae (see p.123) • Sawflies (see p.87)
Large holes, notches, tattered, or entirely eaten	Caterpillars (see p.81) • Swift moths (see p.123) • Slugs and snails (see p.88) • Rabbits (see p.85) • Vine weevils (see p.91) • Sawflies (see p.87)
Small holes	Flea beetles (see p.82) • Capsid bugs (see p.80)
White powder on upper surface	Powdery mildew (see p.95)
Deformed or distorted at shoot tips	Aphids (see p.79) • Capsid bugs (see p.80) • Eelworms (see p.82) • Fasciation (see p.108) • Weedkiller damage (see p.103) • Viruses (see p.99)
Brown marks with fuzzy fungal growth	Downy mildew (see p.93) • Grey mould (see p.94)
Pale or brown tunnels inside leaves	Leaf miners (see p.83)
Turning brown	Eelworms (see p.82) • Frost damage (see p.100) • Drought (see p.100)
Brown or grey spots	Fungal leaf spot (see p.94) • Bacterial leaf spot (see p.92)
Yellowing, sometimes with dry brown patches	Drought (see p.100) • Waterlogging (see p.129) • Viruses (see p.99) • Weedkiller damage (see p.103) Nitrogen deficiency (see p.101) • Magnesium deficiency (see p.1010) • Iron deficiency (see p.101)
Wilting	Foot and root rot (see p.124) • Verticillium wilts (see p.139) • Phytophthora root rot (see p.125) • Root aphids (see p.122) • Vine weevil (see p.123) • Frost damage (see p.100) • Drought (see p.100) • Waterlogging (see p.129)
Small, raised orange or brown pustules	Rusts (see p.96)
FLOWERS AND BUDS	
Brown patches with fuzzy fungal growth	Grey mould (see p.107) • Peony wilt (see p.107)
Deformed, buds may fail to open	Capsid bugs (see p.105) • Gall midges (see p.106) • Fasciation (see p.108)
Black powder on petals	Smuts (see p.139)
Holes or jagged rips in petals	Birds (see p.105) • Earwigs (see p.106) • Slugs and snails (see p.106)
Brown edges or patches on petals	Frost damage (see p.108)
ROOTS	
Fine roots rotted and large roots blackened	Foot and root rot (see p.124) • Phytophthora root rot (see p.125)
Roots severed or eaten	Vine weevil larvae (see p.123)
Small with white waxy coating	Root aphids (see p.122)
Swollen or distorted	Clubroot (see p.124)
Groups of rounded swellings	Crown gall (see p.124)
STEMS	
Groups of swellings	Crown gall (see p.124) • Leafy gall (see p.138)
Wilt and dieback	Drought (see p.140) • Frost damage (see p.140) • Verticillium wilts (see p.139) • Foot and root rots (see p.136) • Phytophthora root rot (see p.125) • Vine weevils (see p.123)
Swollen and split	Eelworms (see p.120) • Smuts (see p.139)

ROSES

Roses display an incredible range of habits, flower colours, and scents, and will produce a profusion of blooms if grown in fertile, free-draining soil. They have a reputation for being prone to pests and diseases, but many problems have little impact on plant growth and can be minimized by choosing the right varieties. Regular feeding, mulching, pruning, and good garden hygiene will help keep your roses in good health.

The thornless variety 'Grace' is perfect for growing in mixed borders.

Floribunda roses, such as 'Iceberg', have some resistance to black spot.

CULTIVATION

Select varieties that are best suited to conditions in your garden and that have resistance to common diseases, such as rose black spot. Plant them in winter or early spring, in moist, well-drained soil that has been improved by forking in plenty of well-rotted compost or farmyard manure. Space the plants according to the final size stated on the label for each variety. Dig a hole twice as wide as the rootball, tease out the roots of container-grown roses, and position so that the bump on the main stem where the variety has been grafted onto the rootstock is above soil level. Firm soil gently around the roots and water well.

PREVENTING PROBLEMS

Roses tend to look best when combined with bulbs, herbaceous perennials, and shrubs for year-round interest. This style of planting is more forgiving than a traditional border filled only with roses because it is easier to tolerate a few marks or nibbles on foliage in a mixed planting. Avoid planting roses where they have been grown before, because pathogens accumulated in the soil can result in a condition known as "rose replant disease", which causes weak growth in new plants. If replanting can't be avoided, remove and replace the topsoil to a depth of 45cm (18in).

Most common pests, diseases, and disorders are readily visible on rose foliage, so watch for spots, yellowing, early leaf fall, or stunted growth, as well as clusters of insects or larvae on young growth. Remove any fallen leaves promptly to prevent fungal spores overwintering or spreading, and feed with a balanced granular fertilizer in early spring, before mulching with well-rotted compost.

Prune to improve airflow within plants, encourage healthy new growth, and promote flowering. Use clean, sharp tools to remove dead, diseased, or damaged stems and then prune as appropriate for the type of rose, using angled cuts just above buds. Cut hybrid tea roses back lightly in autumn to prevent wind rock damaging their roots and tie in climbing and rambling varieties to sturdy supports to prevent the wind breaking their stems.

Disease can be cut out by pruning well below the affected area.

Replace the topsoil if planting a rose on the site of an old rose plant.

COMMON PROBLEMS

Find your rose's problem in the symptoms column, then scan across for possible causes.
Each of these is cross-referenced with a detailed description to help you make the right diagnosis.

SYMPTOMS	POSSIBLE CAUSES
LEAVES	
Insects visible	Aphids (see p.79) • Capsid bugs (see p.80) • Leafhoppers (see p.83) • Red spider mites (see p.86)
Nibbled or eaten entirely	Deer (see p.132) • Rabbits (see p.85)
Pale mottling	Leafhoppers (see p.83) • Red spider mites (see p.86)
Brown patches with fuzzy fungal growth underneath	Downy mildew (see p.93)
Semi-circular notches cut from edges	Leaf-cutting bees (see p.83)
Yellowing and/or wilting, leaves sometimes falling prematurely	Nitrogen deficiency (see p.101) • Magnesium deficiency (see p.101) • Potassium deficiency (see p.102) • Drought (see p.100) • Waterlogging (see p.141) • Rose black spot (see p.96) • Weedkiller damage (see p.102) • Honey fungus (see p.137) • Root aphids (see p.122)
Dark spots	Rose black spot (see p.96)
Leaves tube-like and rolled	Sawfly (see p.87)
White, chalky covering on upper surface	Powdery mildew (see p.95)
Orange spots with raised pustules beneath	Rust (see p.96)
Damaged or deformed at shoot tips	Aphids (see p.79) • Capsid bugs (see p.80) • Viruses (see p.99) • Weedkiller damage (see p.102)
Caterpillars feeding on foliage	Winter moth (see p.91) • Rose slugworm (see p.86)
Yellow or red pompom-like galls	Robin's pin cushion (see p.86)
Yellow patterned marks, stunted growth	Viruses (see p.99)
FLOWERS AND BUDS	
Brown patches with fuzzy fungal growth	Grey mould (see p.107)
Buds brown and fail to open	Rose balling (see p.109)
Insects visible	Aphids (see p.79) • Capsid bugs (see p.80)
Buds stunted, marks or damage on flowers	Aphids (see p.79) • Capsid bugs (see p.80)
Buds deformed, flowers abnormally coloured	Viruses (see p.99) • Weedkiller damage (see p.102)
ROOTS	
Cream or blue-green insects with white waxy covering	Root aphids (see p.122)
Roots above soil and sometimes broken	Wind damage (see p.141)
Fine roots rotten and large roots blackened	Waterlogging (see p.129) • Phytophthora root rot (see p.125)
STEMS	
Rounded swellings at base of main stem	Crown gall (see p.124)
Brown and dry	Dieback (see p.136) • Phytophthora root rot (see p.125) • Honey fungus (see p.137)
Split with bright orange spores	Rust (see p.96)
Vigorous shoots from base with different foliage	Suckers (see p.103)

BULKS

BULBS

Bulbs produce vibrant flowers almost year-round, from yellow aconites in late winter to candy pink nerines in autumn. Given the right conditions, plants reappear healthy and pristine from their underground storage organs (bulbs, corms, tubers, or rhizomes) each year, but watch for wilting, yellowing, or nibbled foliage, indicating pests or diseases that could quickly become serious and spread.

Bulbs like *Tulipa* 'Ballerina' will add that "wow" to your spring garden.

CULTIVATION

Bulbs are made up of stems and leaves that are modified to hold reserves of food and nutrients. It is this store of energy that lets bulbs burst quickly into growth, and gives them a head start against pests and diseases.

Bulbs of woodland species should be planted in a cool, shady spot, while those from hot climates require full summer sun to bloom. All need well-drained soil to prevent rot from spoiling the bulb. Plant spring-flowering bulbs through autumn into early winter, summer-flowering bulbs in autumn or spring, and autumn-flowering types in late summer. Generally, plant bulbs at a depth of roughly three times their height, ensuring that the shoot tips are pointing upwards.

Nerines will form clumps and should be lifted and divided periodically to promote flowering.

PREVENTING PROBLEMS

Always buy your bulbs from a reputable source and check that they are firm and free of mould before planting to avoid introducing fungal diseases. Once planted, be careful not to damage hidden bulbs or new shoots with tools while weeding or digging, and take measures to protect foliage and flowers from slugs and snails where necessary.

After the plants have flowered, let the leaves die down naturally before removing them, because they produce the food stores in the bulb that fuel growth and flowering the following year. If you have planted bulbs within a lawn, you should avoid cutting the grass until early June.

Dahlias, and other bulbs that are not fully hardy, can be lifted and stored in a frost-free place over winter and potted up under cover in spring, before planting out when the weather warms.

Bulbs grown in good conditions are resilient, but look out for signs of pests, diseases, and disorders, and remove any affected plants before problems have a chance to spread. Check for yellowing or wilting foliage, which can be a symptom of fungal diseases or pests at the roots, along with signs of pests feeding on foliage and flowers, as this can rapidly spoil a display and weaken plants for the following season. Large, congested clumps of bulbs can be lifted and divided, and any weak or diseased bulbs removed before replanting.

Alliums are popular bulbs that flower in late spring.

COMMON PROBLEMS

Find your bulb's problem in the symptoms column, then scan across for possible causes, each of which is cross-referenced with detailed directory entries to help you make the right diagnosis.

SYMPTOMS	POSSIBLE CAUSES
LEAVES	
Holes, tears or large portions eaten	Lily beetle (see p.84) • Sawflies (see p.87) • Slugs and snails (see p.88)
Orange or brown pustules	Rusts (see p.96)
Dark spots	Fungal leaf spot (see p.94)
Small yellow spots	Smuts (see p.139)
Yellowing and/or wilting	Foot and root rots (see p.124) • Verticillium wilts (see p.139) • Violet root rot (see p.127) Waterlogging (see p.141)
Young leaves deformed	Aphids (see p.79) • Capsid bugs (see p.80)
White, powdery coating	Powdery mildew (see p.95)
Insects visible	Aphids (see p.79) • Capsid bugs (see p.80) • Lily beetle (see p.84) • Thrips (see p.90) Red spider mites (see p.86)
Pale mottling on surface	Thrips (see p.90) • Red spider mites (see p.86)
Yellow streaks	Viruses (see p.99)
Stunted, withered and distorted	Eelworm (see p.82) • Tulip fire (see p.97) • Viruses (see p.99)
Wilting and blackened or translucent	Frost damage (see p.100)
Failure to grow or very small and weak	Narcissus bulb flies (see p.121) • Mould on bulbs (see p.125) • Tulip fire (see p.97) • Squirrels (see p.122)
FLOWERS AND BUDS	
Holes or tears in petals	Birds (see p.105) • Earwigs (see p.106) • Lily beetle (see p.84) • Slugs and snails (see p.106)
Insects visible, marks or holes on petals	Aphids (see p.105) • Capsid bugs (see p.105) • Lily beetle (see p.84)
Buds fail to open or pale spots on petals	Tulip fire (see p.97)
Bleached streaks or failure to flower	Viruses (see p.107)
Distorted flower stems	Eelworm (see p.120)
ROOTS, TUBERS, AND BULBS	
Roots and bulbs discolour and rot	Foot and root rots (see p.124) • Waterlogging (see p.129)
Bulbs rot with fungal growth or small black growths	Tulip fire (see p.97)
Brown marks on bulb with blue-green growth	Mould on bulbs (see p.125)
Dark purple strands on surface of rotting bulb	Violet root rot (see p.127)
Unearthed and eaten	Squirrels (see p.122)
Grubs inside bulbs and/or roots severed	Vine weevils (see p.123) • Narcissus bulb fly (see p.121)
STEMS	
Growths on stems	Crown gall (see p.124) • Leafy gall (see p.138)
Stems die back and collapse	Verticillium wilts (see p.139) • Frost damage (see p.140)
Raised grey blisters with black powdery spores	Smuts (see p.139)
Bent or broken	Wind damage (see p.141)

BEDDING

Fast-growing bedding plants will fill containers and borders with an abundance of lush foliage and vibrant blooms in a single growing season, but their soft growth is prone to fungal diseases and attack by aphids, slugs, and snails. Hardy bedding plants flower through winter and spring, while half-hardy varieties are damaged by frost and need the warmth of summer to flourish.

Add slow-release fertilizer for a long-lasting profusion of flowers.

Push compost around the plants so no air pockets are left around the roots.

CULTIVATION

Bedding plants can be bought ready to use, or grown from seed or small plug plants on greenhouse staging or on a sunny windowsill. Harden off the seedlings before planting them outdoors; wait until the risk of frost has passed before planting out half-hardy summer bedding. This will give the young plants the best chance to outgrow and withstand pests and diseases.

Choose plants to suit the conditions; pelargoniums and petunias thrive in full sun, while impatiens and begonias do well in shade. Improve soil with well rotted compost and choose multi-purpose compost for containers. Ensure containers have drainage holes and add slow release fertilizer to feed plants consistently.

PREVENTING PROBLEMS

Check young plants before buying, rejecting specimens with visible leaf spots, wilting, or signs of mould. Wherever possible, choose varieties with disease resistance bred in.

Check each plant's final size and avoid planting at too high a density. This will help air to circulate and so keep fungal diseases at bay. Renew the compost in containers each time you replant, and avoid growing the same plants repeatedly in one place to stop a build-up of pests and diseases in the soil.

Keep all bedding plants well watered and improve drainage by adding plenty of organic matter to the soil each year.

Plants in containers or hanging baskets will need daily watering in summer. Install drip irrigation systems in pots for easy maintenance. Raise containers up on "pot feet" during winter to prevent waterlogging of the compost.

Bedding plants will need protection from slugs and snails (see pp.88–89), particularly when newly planted. Regularly check for aphids on young growth and on the undersides of leaves, and squash any that you find. Deadhead plants regularly to prevent them setting seed and keep the flowers coming. Removing fading flowers also stops them from succumbing to fungal infections in wet conditions. Damaged foliage should also be removed for same reason.

Installing a drip irrigation system ensures consistent watering.

Deadhead violas by pinching flowers off between thumb and forefinger.

COMMON PROBLEMS

Find your bedding plant's problem in the symptoms column, then scan across for possible causes, each of which is cross-referenced with detailed directory entries to help you make the right diagnosis.

SYMPTOMS	POSSIBLE CAUSES
LEAVES	
Wilting	Drought (see p.100) • Vine weevil (see p.91) • Waterlogging (see p.141) • Foot and root rots (see p.124) • Clubroot (see p.124) • Cutworms (see p.120)
Insects visible	Aphids (see p.79) • Whitefly (see p.91) • Thrips (see p.90) • Red spider mites (see p.86)
Webbing on foliage	Red spider mites (see p.86)
Powdery white coating	Powdery mildew (see p.95)
Brown edges or tips	Drought (see p.100) • Frost damage (see p.100) • Scorch (see p.103)
Yellowing and/or leaf drop	Nitrogen deficiency (see p.101) • Drought (see p.100) • Viruses (see p.99) • Red spider mites (see p.86) • Root aphids (see p.122)
Water-soaked or blackened	Frost damage (see p.100)
Distorted, often at shoot tips	Aphids (see p.79) • Capsid bugs (see p.81) • Tarsonemid mites (see p.90) • Viruses (see p.99)
Holes or notches in leaves	Vine weevils (see p.91) • Slugs and snails (see p.88) • Flea beetles (see p.82) • Capsid bugs (see p.81)
Brown or grey spots	Fungal leaf spot (see p.94)
Pale mottling on upper surface	Thrips (see p.90) • Red spider mite (see p.86)
Discoloured with fuzzy fungal growth	Downy mildew (see p.93) • Grey mould (see p.94)
FLOWERS AND BUDS	
Brown with fuzzy fungal growth	Grey mould (see p.94)
Brown edges or patches	Frost damage (see p.108) • Wind damage (see p.109)
Insects visible	Aphids (see p.79) • Thrips (see p.90) • Earwigs (see p.106)
Holes or tears in petals	Slugs and snails (see p.106) • Birds (see p.105)
Lack of flowering	Capsid bugs (see p.105) • Potassium deficiency (see p.109)
ROOTS	
Insects feeding among roots	Root aphids (see p.122)
Roots brown or rotted away	Foot and root rots (see p.124) • Waterlogging (see p.129)
Larvae feeding on roots	Vine weevils (see p.123) • Cutworms (see p.120)
Swollen and deformed	Clubroot (see p.124)
STEMS	
Wilting that doesn't recover with watering	Verticillium wilts (see p.139) • Vine weevils (see p.123) • Foot and root rots (see p.124)
Seedlings collapse, fuzzy growth	Damping off (see p.135)
Dieback	Drought (see p.140) • Frost damage (see p.140) • Waterlogging (see p.141)
Damage at base of stem	Cutworms (see p.120)
Wilting that recovers with watering	Drought (see p.140)

LAWNS

A healthy lawn is an inviting place to relax or play and makes the perfect foil for colourful borders. Many gardeners aspire to an immaculate green sward, but some prefer more diversity, leaving daisies and clover to flower, and even introducing wild flowers. Whatever type of lawn you choose, good growing conditions and regular maintenance are essential to avoid problems with moss and fungal diseases.

Traditional lawns use only a limited range of grass species.

Make sure to seed the lawn at the recommended density.

ESTABLISHING A LAWN

You can head off many future problems by taking care to establish a new lawn in the right way. If your soil is heavy, consider taking professional advice on improving drainage to help prevent waterlogging. Remove all weeds from the site, and choose a turf or seed mixture that is suited to the light levels and anticipated wear in your garden.

Sow the seed or lay turf in autumn or spring. Cover newly seeded lawns with netting to protect them from birds and avoid using new lawns heavily for several months while they establish. Mow once the grass reaches a height of 5–7cm (2–3in), and water newly established lawns weekly.

PREVENTING PROBLEMS

There's no need to water established lawns: grass tolerates drought and will green up again as soon as rain falls.

Cut the grass once a week during spring and summer, less often in autumn, and feed traditional lawns with fertilizer from mid-spring to late summer when the soil is moist or rain expected. Apply the fertilizer evenly – overfeeding will "scorch" patches of grass. Use a mechanical spreader for larger lawns.

A dense, healthy lawn will resist weeds, but if they do appear, remove individual weeds by hand or treat larger numbers with a lawn weedkiller that targets the type of weeds in your lawn; always follow directions on the label. Never apply lawn fertilizers or weedkillers to turf containing wild flowers.

Remove fallen leaves and scarify the lawn with a wire rake in autumn to remove moss and dead plant material. Add lawn sand two weeks before scarifying to kill moss and make it easier to remove. Aerate lawns every two or three years, to improve drainage and root health, by pushing a garden fork into turf to a depth of 10cm (4in) at 15cm (6in) intervals. Reduce shade by pruning back overhanging shrubs.

Scarifying your lawn gets rid of "thatch", improving air circulation around the plants, which helps prevent fungal diseases.

COMMON PROBLEMS

Find your lawn's problem in the symptoms column, then scan across for possible causes, each of which is cross-referenced with detailed directory entries to help you make the right diagnosis.

SYMPTOMS	POSSIBLE CAUSES
LEAVES	
General yellowing over whole lawn	Drought (see p.100) • Waterlogging (see p.141) • Nitrogen deficiency (see p.101)
Pale yellow or straw-coloured patches	Turf fungal diseases (see p.98) • Weedkiller damage (see p.103) • Chafer grubs (see p.120)
Toadstools in turf	Fairy rings or turf toadstools (see p.98)
Large concentric rings of lush then dying grass	Fairy rings (see p.98)
Thin, pink-red fungal strands in yellowed patches	Red thread (see p.98)
Fuzzy, pale pink growth in yellowed patches	Snow mould (see p.98)
Dense, white fungal growth in yellowed patches	Turf thatch fungal mycelium (see p.98)
Orange or black pustules on yellowed leaves	Turf rust (see p.98)
Holes scratched in turf by birds and mammals searching for insect larvae	Chafer grubs (see p.120) • Leatherjackets (see p.121)
Poor growth, dead patches, moss	Waterlogging (see p.141)
Brown patches following line of footprints on frozen grass	Frost damage (see p.100)
ROOTS	
Insect larvae feeding on roots	Chafer grubs (see p.120) • Leatherjackets (see p.121)

ANIMAL ISSUES

Wild animals and pets can damage your lawn. If you're unlucky, you may have active moles in your area. Try to tolerate their activity: these animals do have some benefits in the garden (see p.28) and killing them is unnecessary.

Worm casts can be brushed off into the lawn in dry weather.

Remove and flatten loose molehill soil before mowing, and reseed bare patches in spring or autumn. Moles are deterred by noisy activity, like mowing, children playing, or buzzing deterrents. Smoke repellants may discourage the animals, but nearby moles may simply move into the unused tunnels later. Soil brought up by worms can easily be brushed back into turf.

From autumn to spring, birds, foxes, and badgers may dig up areas of turf in search of chafer grubs: just level and reseed to repair the damage. Regular lawn maintenance and the application of nematode biological control (see pp.42–43) in late summer should limit damage by reducing grub numbers.

Dog urine causes brown patches of grass surrounded by lush growth. Try to keep dogs off lawns where possible and water affected areas immediately to reduce damage.

Solar-powered mole deterrents emit sound into the ground.

Diagnosing the cause of leaf spots and other symptoms quickly becomes easier with experience and helps to determine whether damage can be tolerated or if action is needed.

DIRECTORY OF PESTS, DISEASES, AND DISORDERS

Packed with illustrated entries describing the common pests, diseases, and disorders that trouble garden plants, this chapter will enable you to identify the cause of symptoms and choose the most effective method of treatment or control. For quick reference, symptoms are listed under the part of the plant affected – leaves, flowers and buds, fruits and pods, roots, tubers, and bulbs, and stems.

Brunnera macrophylla **'Jack Frost'** leaves are highly decorative and tough enough to resist attacks by most pests and diseases.

LEAVES

Healthy foliage is integral to the survival of every plant, since the leaves are the site of photosynthesis. Pests may feed on entire leaves, or create holes, notches, tunnels, or deformities at shoot tips. Diseases often result in characteristic marks on foliage, ranging from flat or raised spots and blotches to mottled discolouration. The condition of the leaves can be also an indicator of general health because leaves may wilt or change colour when water or nutrients are scarce, or if roots and stems are under attack.

APHIDS APHIDOIDEA

Aphids are small, sap-sucking insects that feed on a huge range of ornamental and edible plants, especially during mild weather. The most common ones found in gardens are greenfly or blackfly (general names for various green and black species), but aphids may be pink, yellow, or brown.

PLANTS AFFECTED Many plants outdoors and in greenhouses

SEASON Spring to autumn

WATCH FOR Insects clustered at shoot tips, deformed leaves, honeydew and sooty mould

ACTION NEEDED? Yes, for serious cases

SEE ALSO Pages 105, 131, 133, and 122 for aphid damage to flowers and buds, stems, and roots

SYMPTOMS

Aphids are so prevalent that you have little choice but to tolerate them to a degree. Plants usually recover well from light infestations, but aphids can multiply incredibly quickly on young growth and cause serious damage if unchecked.

Aphids will affect plant vigour and cause young leaves to become deformed, curled, or puckered. Clusters of insects are visible on shoots or the underside of leaves, along with papery skin casts left by growing nymphs. Some aphids excrete sweet, sticky honeydew onto foliage, which may attract ants and turn black with sooty mould. Many plants shrug off damage, but young plants, vegetable crops, and the new foliage of fruit trees can be seriously affected. Warm conditions in greenhouses allow aphids to remain active over a longer season.

PREVENTION

Encourage aphid predators, including ladybirds, hoverflies, birds, and earwigs, into your garden (see pp.30–31) to feast on colonies throughout the growing season. It will take time for them to start reducing aphid numbers, because the predator population must build up; their activity should be keeping aphids in check by summer. Prevent blackfly from infesting broad bean plants by pinching out the beans' growing tips in late spring and grow nasturtiums as sacrificial plants (see p.35) to draw aphids away from French and runner beans.

CONTROL

Promptly squash any aphids you see on leaves, flowers, and stems. Kill eggs overwintering on the bark of fruit trees by applying a winter wash containing plant oils. Several biological controls are available for aphids in greenhouses (see pp.42–43).

All available chemical controls will kill aphid predators as well as aphids, so use them only when absolutely necessary. Sprays with natural pyrethrum, plant oils, and fatty acids are effective. Insecticides with the synthetic pyrethroids lamba-cyhalothrin, deltamethrin, and cypermethrin work on contact, while the neonicotinoid acetamiprid is systemic and will reach aphids under curled leaves.

Blistering on foliage, as on this currant plant, may be caused by aphid damage.

A single ladybird may consume more than 50 aphids a day. Ladybird and lacewing larvae can be bought for release in your garden.

ASPARAGUS BEETLE *CRIOCERIS ASPARAGI*

PLANTS AFFECTED Asparagus
SEASON Spring to autumn
WATCH FOR Defoliation and bark damage
ACTION NEEDED? Yes, to control beetle numbers

These distinctive, colourful beetles and their grubs feed on the leaves and bark of asparagus plants. Areas above where bark has been eaten around the stem turn brown and dry out, and significant damage to plants can reduce the crop the following spring. Remove beetles and grubs by hand to control numbers. Cut down and burn stems in late autumn to destroy overwintering beetles inside. Serious infestations can be treated with organic pesticides containing pyrethrum.

Asparagus beetles are red and black and have yellow spots on their backs.

BEAN SEED FLY *DELIA PLATURA*

PLANTS AFFECTED French and runner beans
SEASON Spring and summer
WATCH FOR Weak, damaged seedlings
ACTION NEEDED? Some, to reduce risk of attack

Seedlings of French and runner beans affected by this pest emerge with tatty leaves and stem damage and grow slowly, or may fail to emerge at all. This damage is caused by white, soil-dwelling fly larvae feeding on germinating beans. There are no methods of control. Reduce the risk of attack by applying organic matter to beds in autumn rather than spring, avoiding sowing seeds in cold, wet soil where slow germination will make them vulnerable, or by sowing in pots, to plant out once leaves are fully developed.

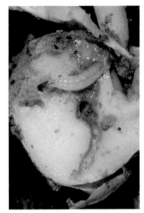

Bean seed fly larvae are 8mm (⅓in) long and hatch from eggs laid in the soil.

BOX / BAY SUCKERS *PSYLLA / LAURITRIOZA* SPECIES

PLANTS AFFECTED Box and bay
SEASON Spring and summer
WATCH FOR Leaves curl at margins, thickening and yellowing on bay
ACTION NEEDED? No, damage can usually be tolerated

These small bugs and their nymphs suck sap from young foliage, causing it to curl at the edges. On box, leaves become cupped and marked with white, waxy drops. Sucker damage does not affect vigour and is often removed during clipping. Bay leaves can also be picked off individually. Chemical controls are unlikely to be effective on bay. Heavy infestations can be treated with organic pyrethrums, fatty acids, or plant oils, or contact insecticides with synthetic pyrethroids.

Affected bay leaf edges curl, thicken, and yellow; white, waxy insects may be seen below.

CAPSID BUGS *MIRIDAE*

PLANTS AFFECTED Many shrubs, climbers, perennials, and vegetables
SEASON Spring and summer
WATCH FOR Deformed leaves, peppered with holes at shoot tips
ACTION NEEDED? Sometimes, in known areas of activity

These 6mm (¼in) long bugs suck sap from the shoot tips of a wide range of plants. The small holes they make expand as the leaves grow, giving foliage a tattered look. Encourage beneficial wildlife to prey on these bugs and keep beds free of weeds and dead plant material where bugs can overwinter. Where insects or early damage are seen, treat with organic pyrethrum, fatty acid, or plant oil sprays, or contact insecticides with lambda-cyhalothrin, deltamethrin, or cypermethrin.

Once capsid bug damage becomes visible, it's too late to take any effective action.

CATERPILLARS LEPIDOPTERA

Caterpillars are the larvae of butterflies and moths. They hatch from eggs usually laid on the leaves of a suitable food plant. There are many species which feed on a huge variety of garden plants. Most can be tolerated to some degree and controlled using similar methods if necessary.

PLANTS AFFECTED A broad range of edible and ornamental plants
SEASON Spring to autumn
WATCH FOR Holes in leaves or defoliation, silk webbing, feeding larvae
ACTION NEEDED? Sometimes, where damage is severe

SYMPTOMS

You will often observe caterpillars directly on plant leaves. Some are camouflaged in shades of green and brown to avoid predators; others are patterned with bold stripes, or covered with hairs that serve as a deterrent. Typical signs of activity are holes in leaves or defoliation, where only the central rib of each leaf remains. Silk webbing, produced by some caterpillars as a defence, may also be evident.

Caterpillar damage in early spring is likely to be caused by the winter moth (see p.91). Larvae of the lackey moth and vapourer moth feed on a range of trees and shrubs from late spring into summer. Otherwise, larvae are specific to host plants: box tree moth caterpillars eat only box leaves; colourful mullein moth caterpillars feed only on mulleins, figworts, and buddlejas; plants in the cabbage family and nasturtiums are eaten by caterpillars of large and small white butterflies and cabbage moth.

PREVENTION

The presence of caterpillars should be tolerated as much as possible, as they provide a valuable food source for many birds, parasitic wasps, and beetles, which help to keep numbers under control and reduce damage to plants. Prevent eggs being laid on vegetable crops by protecting them with fine netting supported on a frame, so that eggs can't be laid onto leaves through it.

CONTROL

Pick caterpillars off plants by hand and squash any eggs found on the undersides of leaves. Pheromone traps (see p.35) can be useful to check for the presence of egg-laying adults and monitor their numbers. In greenhouses, try the biological control *Trichogramma brassicae* – a wasp that targets caterpillar eggs. Such controls are also available for use on box tree moth and cabbage caterpillars. Chemical controls can be used when necessary. Organic pyrethrins and the synthetic pyrethroids lambda-cyhalothrin, deltamethrin, and cypermethrin are all effective.

The mullein moth lays eggs on verbascum, buddleia, and figwort.

Brassicas should be grown under netting to protect them from cabbage whites, which may lay two or three times in summer.

EARWIGS *FORFICULA AURICULARIA*

PLANTS AFFECTED Dahlias, chrysanthemums, and clematis
SEASON Late spring to autumn
WATCH FOR Ragged holes in young leaves and flowers
ACTION NEEDED? Sometimes, to control numbers

Earwigs are brown, crawling insects, easily recognized by the pair of pincers at their rear. Adults feed at night on a variety of insects including aphids, but also by nibbling holes in young flowers and leaves, which then expand as they grow. Tolerate damage, or pick earwigs off plants after dark. They will hide in flower pots loosely stuffed with straw, placed upside down on canes, from where they can be collected each day. Avoid using chemical sprays – earwigs are active when plants are in flower.

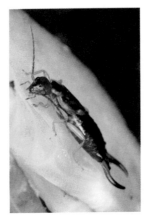

Earwigs can be tolerated, especially on fruit trees, where they help control aphids.

EELWORMS NEMATODA

PLANTS AFFECTED Wide range, including vegetables and perennials
SEASON Summer and autumn
WATCH FOR Brown patches, distorted and stunted growth, dieback
ACTION NEEDED? Yes, to prevent spread

Many species of these microscopic worm-like animals live in garden soil. Most are harmless, but some are serious plant pests, which can also transmit viruses. Eelworms that infect leaves cause brown patches bordered by leaf veins. Root eelworms make foliage die back from the base of a plant or cause weak and deformed growth. Avoid wetting foliage to help prevent eelworms spreading. Choose resistant varieties where available. No effective chemical controls are available.

Infected leaves will dry up and die. Dispose of infected plants by burning. Do not compost.

FLEA BEETLES *PHYLLOTRETA / PSYLLIODES* SPECIES

PLANTS AFFECTED Edible and ornamental brassicas, fuchsias
SEASON Spring and summer
WATCH FOR Small round holes in leaves
ACTION NEEDED? Sometimes, to prevent serious damage

These tiny black beetles may be seen jumping when disturbed. They eat rounded holes in the leaves of fuchsias and brassicas. Light damage is tolerable, but heavy feeding can kill seedlings and affect plant growth. Grow susceptible crops under fine insect-proof mesh and control heavy infestations with a contact insecticide. Organic sprays containing natural pyrethrins are effective, as are the synthetic pyrethroids deltamethrin, lambda-cyhalothrin, and cypermethrin.

Reduce damage by watering seedlings well so that they rapidly outgrow danger.

GALL MITES ERIOPHYIDAE

PLANTS AFFECTED Trees, shrubs, and fruit bushes
SEASON Spring and summer
WATCH FOR Blisters, deformed growth, galls
ACTION NEEDED? No

These mites are too small to see with the naked eye but the chemicals they secrete when feeding on plants cause distinctive growths that are hard to miss. These include leaf blisters, felty or spiny growths on leaves, thickening and curling of leaf margins, distortion of new shoots, or enlarged leaf buds that fail to open or develop into galls. Despite their strange appearance, these symptoms rarely affect plant growth or general health and need to be tolerated as no treatments are available.

Affected plants, such as this acer, can be selectively pruned to improve their appearance.

LEAF-CUTTING BEES MEGACHILIDAE

PLANTS AFFECTED Roses and sometimes other plants
SEASON Summer
WATCH FOR Semi-circular sections cut from leaf edges
ACTION NEEDED? No

Female solitary bees, which resemble honeybees, will visit certain plants from which they cut sections of leaves to build their nests. Roses are often affected, but bees will also harvest foliage from epimediums, wisterias, and many other plants. The elliptical or circular leaf portions are taken from leaf margins and the cut edges are smooth, unlike damage by other leaf pests. The leaf area lost is not sufficient to set back growth. The bees are also pollinators so their presence can be tolerated.

These rose leaves show the characteristic rounded holes made by leaf-cutting bees.

LEAFHOPPERS CICADELLIDAE

PLANTS AFFECTED Many trees, shrubs, and perennials
SEASON Spring and summer
WATCH FOR Pale mottling on foliage
ACTION NEEDED? Sometimes, for severe infestations

Pale stippled marks appear on upper leaf surfaces. They are caused by small yellow or pale green insects, about 2mm (1/12in) in length, sucking sap from the underside of the leaves. Adults can sometimes be seen jumping from leaves if disturbed, while the cream-coloured nymphs remain on foliage. If necessary, chemical sprays containing organic pyrethrums, fatty acids, or plant oils will provide control, as will contact insecticides containing cypermethrin and deltamethrin. Do not spray plants in flower.

A sage leaf displays moderate leafhopper damage, which can usually be tolerated.

LEAF MINERS VARIOUS SPECIES

PLANTS AFFECTED Many edible and ornamental plants
SEASON Late spring to autumn
WATCH FOR Pale tunnels or patches within leaves
ACTION NEEDED? Control to prevent infestation

Insect larvae, including those of some flies and moths, may feed by mining tunnels inside leaves. Mined areas appear as pale patches or pathways, which eventually turn brown. Larvae may be visible within. Extensive tunnelling may cause leaves to die and fall; established plants can cope, but younger ones may be set back. Remove affected leaves or squeeze tunnels to squash the larvae inside. Grow beetroot and Swiss chard under insect-proof mesh to prevent insects laying eggs, and rotate crops each year.

Leaf miner damage is very distinctive. No chemical controls are available.

LEEK MOTH ACROLEPIOPSIS ASSECTELLA

PLANTS AFFECTED Leeks, onions, shallots, and garlic
SEASON Spring to autumn
WATCH FOR Pale or brown marks on leaves, silk cocoons
ACTION NEEDED? Yes, protect crops where pest present

White or brown patches appear on foliage where leek moth caterpillars have tunnelled while feeding. Tunnels extend into stems and bulbs, allowing in secondary rots. Caterpillars emerge and pupate on the outside of the plant in pale, silky cocoons. Damage can kill young plants and spoil crops. Protect vulnerable plants under insect-proof mesh in spring and summer to prevent moths landing to lay eggs. Crush any cocoons found on foliage. No chemical controls are available.

The leek moth caterpillar grows to 12mm (1/2in) in length. It is pale with a brown head.

LILY BEETLE *LILIOCERIS LILII*

PLANTS AFFECTED Lilies and fritillaries
SEASON Spring to autumn
WATCH FOR Damaged foliage and bright red beetles
ACTION NEEDED? Some, to control numbers

These distinctive bright red beetles, around 8mm (⅓in) in length, can be spotted eating rounded holes in the leaves, petals, and seedpods of lilies and fritillaries, between early spring and autumn. They lay clusters of elongated, orange eggs on the underside of leaves from late spring into summer. These hatch into larvae that appear shiny and black because they cover themselves in their own excrement. Newly hatched larvae feed on the underside of leaves, causing brown patches to develop, but move on to eating entire leaves, flowers, and seedpods, spoiling the appearance of plants and sometimes preventing bulbs swelling for the following year's growth.

CONTROL Small numbers of lily beetles can be tolerated. Their effects can be reduced significantly simply by removing conspicuous beetles and grubs from plants by hand. Encourage beneficial creatures, like birds and frogs, into your garden to feed on the larvae. All chemical controls are more effective against larvae than adults and should only be used in spring,

Lily beetles are among the most serious pests of lilies and tree lilies, capable of devastating a plant within a few days.

before plants flower, to avoid harming pollinating insects. Organic sprays containing natural pyrethrums, as well as insecticides including the synthetic pyrethroids deltamethrin, lambda-cyhalothrin, and cypermethrin and the neonicotinoid acetamiprid are effective.

MINT BEETLE *CHRYSOLINA HERBACEA*

PLANTS AFFECTED Mint
SEASON Summer
WATCH FOR Holes in leaves and blue or green beetles
ACTION NEEDED? Perhaps, to control numbers

These glossy dark blue or green beetles, around 7mm (⅓in) in length, eat holes in the leaves of mint plants. Their black grubs reach about 1cm (⅖in) in length and also feed on foliage. The damage caused may be unattractive, but rarely affects the plant's health, so can usually be tolerated. Adult beetles and grubs are easy to pick off during summer and will be preyed upon by birds, frogs, and other wildlife. Treat heavy infestations with natural pyrethrum to kill the larvae.

Mint is such a prolific plant that it is rarely necessary to control the beetle pest.

PEA AND BEAN WEEVIL *SITONA LINEATUS*

PLANTS AFFECTED Peas and broad beans
SEASON Spring
WATCH FOR U-shaped notches in leaf margins
ACTION NEEDED? No

Regular notches eaten into the edges of pea and broad bean leaves are an obvious sign that a pest is feeding. The small, grey-brown weevils responsible can be hard to spot as they drop to the ground when disturbed. The larvae feed below ground on the plant's nitrogen-fixing root nodules. Established plants will be unaffected but the growth of seedlings may be checked. However, they will generally recover, especially if early sowings are covered with horticultural fleece to protect them from the cold.

Young plants are affected in spring by adult weevils that overwintered in leaf litter.

PEAR LEAF BLISTER MITE *ERIOPHYES PYRI*

PLANTS AFFECTED Pear trees
SEASON Spring to autumn
WATCH FOR Pale or pink, then black, blisters on leaves
ACTION NEEDED? No

The appearance in spring of slightly raised pale yellow or pink leaf blisters may indicate the activity of this mite. These marks are typically in broad lines either side of the central vein of each pear tree leaf and gradually darken to black in summer. They are caused by chemicals released by the microscopic mites as they feed within the leaves. The damage does not affect growth or fruit crops. Where only a few leaves are affected they can be removed, but otherwise this pest has to be tolerated.

The mites that feed on pear leaves are so small as to be almost invisible to the naked eye.

PIGEONS COLUMBIDAE

PLANTS AFFECTED Brassica crops, peas, and some ornamentals
SEASON All year round, but especially winter
WATCH FOR Jagged pecks in foliage
ACTION NEEDED? Yes, particularly for edible crops

Wood pigeons are large birds with appetites to match. They graze on the foliage of vegetables from the cabbage family (sprouting broccoli and kale) as well as pea shoots, often stripping the leaves of mature plants, or seriously checking the growth of seedlings. They will also feed on the buds and fruits of cherries and currant bushes, and lilac trees. Deterrents are unlikely to be effective, so keep pigeons away from vegetable and fruit crops by covering them securely with netting.

Pigeons love brassicas, and can quickly strip a plant down to bare stalks.

RABBITS LEPORIDAE

PLANTS AFFECTED Wide range of edibles and ornamentals
SEASON All year round
WATCH FOR Whole plants grazed, or damage to soft shoots and bark on woody plants
ACTION NEEDED? Yes

Rabbits can cause serious problems for gardeners at any time of year, because they eat a huge range of bedding plants, bulbs, perennials, and vegetable plants down to ground level, along with any soft growth they can reach on shrubs. They will also gnaw the bark of woody plants, particularly in winter, which will kill trees and shrubs if the damage extends around the trunk. Even a single rabbit can wreak havoc in a garden.

CONTROL It is not feasible to trap or kill rabbits in gardens and if there are populations on adjoining land this would be a pointless task in any case. Exclude rabbits from flower beds or vegetable patches by enclosing them with strong netting at least 90cm (3ft) high, supported on sturdy poles and secured at the base. Protect young trees by covering the base of their trunks with plastic tree guards. A rabbit-proof fence can also be erected around the perimeter of a garden. This must be made of wire mesh with holes no wider than 2.5cm (1in), to

Tender young leaves are a rabbit's preferred food, but they will consume stems, fruit, vegetables, and more.

prevent baby rabbits squeezing through, and should be at least 1.2m (4ft) high, with 30cm (12in) of mesh below soil level, to stop rabbits tunnelling underneath. Check neighbouring gardens to see what survives rabbit attacks in your area or try growing plants that they are reputed to find unappealing, including peonies, buddlejas, and rosemary.

RED SPIDER MITES *TETRANYCHUS* SPECIES

PLANTS AFFECTED Wide range of edibles and ornamentals
SEASON Summer outdoors, year-round in greenhouse
WATCH FOR Pale mottling on leaves, silky webbing
ACTION NEEDED? Yes, particularly in the greenhouse

These tiny, sap-sucking mites, less than 1mm (1/24in) long, are more usually yellow-green or dark brown. They thrive in warm, dry conditions, causing pale mottling on leaf surfaces, early leaf fall, and silk webbing on badly affected plants. Avoid overcrowding greenhouses and discourage mites by misting plants in summer to increase humidity. Organic sprays with natural pyrethrum, fatty acids, or plant oils provide some control, as do biological controls. Fumigants based on garlic extract can be used in greenhouses.

Grape vine foliage shows the mottling caused by red spider mites.

ROBIN'S PIN CUSHION *DIPLOLEPIS ROSAE*

PLANTS AFFECTED Wild and species roses
SEASON Summer and autumn
WATCH FOR Pompom-like growths on stems
ACTION NEEDED? No

Spherical growths covered with soft yellow or red tufts may appear on the stems or foliage of wild and species roses. These are galls – protective structures produced by the plant when a gall wasp lays eggs in its developing buds. White larvae hatch, feed, grow, and overwinter inside the galls and then pupate in spring. Although they look alarming, the galls don't harm their host and control measures are unnecessary. Affected stems can be pruned out if they spoil the appearance of the plant.

A red, fibrous growth on rose stems indicates the activity of gall wasps.

ROSEMARY BEETLE *CHRYSOLINA AMERICANA*

PLANTS AFFECTED Rosemary, lavender, thyme
SEASON All year
WATCH FOR Striped beetles, damaged foliage and flowers
ACTION NEEDED? Some, to control numbers

These iridescent green- and purple-striped beetles occur year-round on Mediterranean herbs. They lay eggs in late summer, when the adults and resulting grey larvae begin feeding on leaves and flowers. Natural pyrethrins and the synthetic pyrethroids deltamethrin, lambda-cyhalothrin, and cypermethrin provide contact control, especially of larvae, while the neonicotinoid acetamiprid has a systemic action. Do not spray plants in flower and only use permitted pesticides on edible herbs.

Rosemary beetle also affects lavender. Remove adults by hand when you see them.

ROSE SLUGWORM *ENDELOMYIA AETHIOPS*

PLANTS AFFECTED Roses
SEASON Summer and autumn
WATCH FOR Caterpillar-like larvae, pale patches on leaves
ACTION NEEDED? Some, to control numbers

These pale yellow-green caterpillar-like slugs are the larvae of a sawfly. They feed on the underside of rose leaves, causing the foliage to turn pale, dry, and brown. Watch for symptoms from early summer and pick larvae off leaves by hand. Roses bloom when slugworms are active, which prevents the use of chemical controls, but where plants are not in flower, natural pyrethrins and the synthetic pyrethroids deltamethrin, lambda-cyhalothrin, and cypermethrin are effective.

Slugworms are around 10mm (3/8in) long. Their action has little impact on plant growth but can be unsightly.

SAWFLIES SYMPHYTA

PLANTS AFFECTED Gooseberries, Solomon's seal, various fruit trees, shrubs, and perennials

SEASON Spring and summer

WATCH FOR Caterpillar-like larvae, rapid defoliation, dark holes on developing fruit

ACTION NEEDED? Yes

Sawfly larvae feed voraciously on the foliage or fruit of many plants, before dropping to the soil to pupate. The small, pale grey or green caterpillar-like larvae often have dark heads and black spots on their bodies and are capable of stripping all the foliage from a plant in two or three weeks. Rose leaf-rolling sawfly causes foliage to curl over to protect its eggs and feeding larvae, and tends to be less destructive. Sawfly larvae may burrow into the developing fruits of apple and plum trees, leaving messy holes, or tunnel near the surface, often causing fruitlets to drop.

CONTROL Check vulnerable plants regularly through spring and summer for newly hatched larvae, picking them off by hand to limit damage. Remove rolled leaves from roses and collect fallen fruitlets from beneath trees to help prevent another generation of flies laying eggs. Encourage natural predators

Gooseberry sawfly larvae eat leaves from the leaf margins towards the midrib. A heavy infestation can strip a bush in days.

such as birds and ground beetles into the garden. Nematode biological controls that are watered on to affected plants can help to control an infestation. Contact insecticides containing natural pyrethrins will kill larvae, as will those formulated with the synthetic pyrethroids deltamethrin, lambda-cyhalothrin, and cypermethrin, but check their suitability for use on edible plants.

Scale insects around 2–6mm (¹⁄₁₂–¼in) lay eggs under a white, waxy protective layer (visible above) in spring.

SCALE INSECTS COCCOIDEA

PLANTS AFFECTED Trees, shrubs, woody climbers

SEASON All year

WATCH FOR Small bumps on stems or under leaves

ACTION NEEDED? Yes, to prevent infestation

Hidden under protective domed, brown, yellow, or grey shells, these sap-sucking insects can be hard to spot, especially as they cluster on lower leaf surfaces and on bark. They can be tolerated on large plants but younger plants may be weakened, and the sticky honeydew that some species excrete can cause upper leaf surfaces to be blackened with sooty mould.

CONTROL Small clusters of scale insects can be picked off by hand. Natural predators, such as birds and ladybirds, may help control numbers, and a nematode biological control is available to treat some species. Chemical controls are largely ineffective against adults, so spray in midsummer when the nymphs have hatched. Organic sprays containing natural pyrethrums, fatty acids, and plant oils are all effective against nymphs, as are plant invigorators with surfactants or fatty acids. Contact insecticides, including the synthetic pyrethroids deltamethrin, lambda-cyhalothrin, and cypermethrin, and systemic insecticides containing the neonicotinoid acetamiprid are also effective.

SLUGS AND SNAILS

GASTROPODA

The number one pest in many gardens, these molluscs munch rapidly through plant tissue using their rasping mouth parts. The common garden snail, with its flecked brown shell, is active between spring and autumn, while a variety of slug species feed during mild weather throughout the year.

PLANTS AFFECTED Many plants
SEASON All year
WATCH FOR Irregular holes in leaves
ACTION NEEDED? Yes, to prevent serious damage

SEE ALSO Pages 106, 122, and 133 for slug and snail damage to flowers and buds, roots and tubers, and stems

Fresh bite marks have ragged edges; older ones appear rounded and yellowed.

SYMPTOMS

The surest signs of slug and snail activity are untidy holes that appear overnight in the centre of leaves, particularly in mild, damp conditions. Any soft foliage is vulnerable to attack, especially the new spring shoots of perennials, seedlings, and young vegetable plants. Slug and snail damage during these critical early stages of growth can be devastating and young plants may be killed or seriously set back. While established plants are more resilient, foliage shredded by slug and snail attack ruins their appearance and makes edible crops unappetising.

PREVENTION

The best policy is to protect your plants and reduce slug and snail populations before they become a problem. Encourage hedgehogs, shrews, toads, beetles, and other predators into your plot to keep numbers down (see pp.30–31). Slugs and snails love to shelter in shady hiding places during the day, so eliminate potential sanctuaries. Remove piles of stones or bricks, tidy up leafy debris, and don't store empty pots in the greenhouse.

Many gardeners choose to protect plants using barriers to slug and snail movement, such as copper tape, egg shells, and pine needles. These are not proven to be effective on their own but may help as part of a broader pest control strategy.

Where small vegetable seedlings are being devoured by slugs and snails, try raising the seedlings in containers elevated on a rack or table, and plant them out when they are larger and more robust. Protect young plants with cloches (you can make your own from clear plastic bottles with their bases cut off) while they establish. Creating a boundary of sacrificial plants, like lettuce, may cause a distraction for long enough to allow your desired crop to succeed. If all else fails, grow plants with aromatic, hairy, tough, or glossy leaves, which tend to be ignored.

Copper tape around a pot may help to deter slugs and snails.

Clear plastic water bottles can be repurposed as protective cloches.

A common garden snail leaves behind a trail of mucus. This slimy substance helps the snail stick to and move across its substrate, and protects its exposed soft parts.

CONTROL

Slugs and snails are active at night, so mount torchlit patrols after dark in spring and summer to collect and remove as many as possible. Late in the year, look for sheltered spots where snails gather to overwinter and remove them too. Create an attractive hiding place by propping the skin of half a grapefruit on a stone among plants, then empty it of gastropods each morning. Beer traps, made by sinking a jar one-quarter filled with beer into the soil, are also a magnet for slugs and snails, which drown when they fall into the liquid. A garlic drench (see p.49) regularly sprayed over vulnerable plants is often an effective deterrent.

Nematode biological controls reduce slug populations by infecting them with a fatal bacterial disease, but have less impact on snails and need to be applied in warm, moist conditions several times a year. Chemical controls are available in the form of slug pellets, which should be scattered thinly around plants according to the label directions. Pellets containing ferric phosphate are approved for organic growers, while those containing metaldehyde are not, and their toxicity presents a considerable risk to garden wildlife.

Beer traps are effective in spring but need to be emptied and replenished every day.

TARSONEMID MITES TARSONEMIDAE

PLANTS AFFECTED Mainly greenhouse plants
SEASON Summer and autumn outdoors, year-round in greenhouses
WATCH FOR Distorted flowers and stunted growth
ACTION NEEDED? Yes, to prevent spread

These tiny, sap-sucking mites feed on the flowers and shoot tips of greenhouse plants such as pelargoniums and cyclamen and outdoor plants including Michaelmas daisies and strawberries. Their feeding slows growth and can distort young leaves and flowers. Serious infestations may result in brown marks on stems and growth stopping altogether. There are no chemical controls. Remove affected plants immediately or use predatory *Amblyseius* mites as a biological control in greenhouses.

The mites are not very mobile and are often brought in on infected plants.

THRIPS THYSANOPTERA

PLANTS AFFECTED Wide range of edibles and ornamentals
SEASON Spring to autumn
WATCH FOR Pale, silvery mottling on leaves
ACTION NEEDED? Sometimes, if damage is severe

Thrips are slender insects, about 2mm (1/12in) long, that suck sap from the upper leaf surface of many plants. Leaves lose their lustre and develop fine, pale markings, and shoots and flower buds become distorted. Treat if growth is affected. Biological controls are available for use in greenhouses, where sticky traps and fumigants can also control numbers. Outdoors, try organic sprays containing pyrethrum, fatty acids, or plant oils, or contact insecticides containing synthetic pyrethroids.

Thrip damage can be tolerated, but the insects are also vectors of plant viruses.

TORTRIX MOTH TORTRICIDAE

PLANTS AFFECTED Many garden and greenhouse plants
SEASON Spring and summer outdoors, all year in greenhouses
WATCH FOR Silky webbing concealing small, green caterpillars
ACTION NEEDED? Sometimes, to control damage

The caterpillars of this moth create shelters by binding leaves together with fine, white webbing. They feed on the inner surface of the leaves, causing damaged areas to turn brown. Crush affected leaves between thumb and forefinger and use pheromone traps to monitor numbers of adult moths. The caterpillars are difficult to reach with contact insecticides: very thorough spraying with natural pyrethrum or synthetic pyrethroids is needed for effective treatment.

A tortix moth caterpillar is visible here surrounded by its white webbing.

VIBURNUM BEETLE PYRRHALTA VIBURNI

PLANTS AFFECTED Viburnums, particularly V. tinus and V. opulus
SEASON Spring and summer
WATCH FOR Holes eaten in foliage until only leaf veins remain
ACTION NEEDED? Sometimes, to prevent serious defoliation

From mid-spring until early summer, small cream larvae with black spotted markings feed on the young leaves of *Viburnum* shrubs. Small brown adult beetles also eat foliage from summer into autumn, but cause less damage. Encourage birds into your garden to prey on larvae and adults. Remove larvae by hand in spring or use organic sprays containing natural pyrethrins, contact insecticides with synthetic pyrethroids, or sprays containing the neonicotinoid acetamiprid.

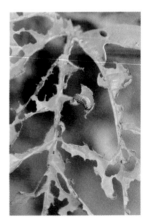

Larvae chew through leaves, making them ragged, and often leaving little behind.

VINE WEEVIL (ADULTS) *OTIORHYNCHUS SULCATUS*

PLANTS AFFECTED Many, including rhododendrons and hydrangeas
SEASON Spring to autumn
WATCH FOR Notches eaten from leaf edges
ACTION NEEDED? Rarely where only leaves are damaged

Adult vine weevils are matt black beetles, 1cm (⅖in) long, that eat irregular notches into the leaf margins of shrubs and herbaceous plants. This rarely affects growth, but signals that grubs might be causing more serious damage to the roots of nearby plants (see p.123). Shake beetles off plants on to newspaper after dark on warm evenings, or pick them off by hand. Biological controls are also available: these employ a parasitic nematode to infect and control adult beetles.

Vine weevils can multiply quickly because females do not need males to reproduce.

WHITEFLY ALEYRODIDAE

PLANTS AFFECTED Greenhouse plants, outdoor shrubs, and brassicas
SEASON All year
WATCH FOR White insects, sticky honeydew and sooty mould
ACTION NEEDED? Yes, when found in the greenhouse

These small, white-winged insects suck sap from the underside of leaves and fly up if disturbed. Their sticky honeydew coats leaves, which can develop black sooty mould. Infestations may occur in greenhouses, where the wasp *Encarsia formosa* is used as a biological control. Organic sprays with natural pyrethrums, fatty acids, and plant oils provide control, as do contact insecticides containing synthetic pyrethroids, and systemic products with the neonicotinoid acetamiprid.

Cabbage whitefly is one of more than 1400 species of this insect pest.

WILTING VARIOUS CAUSES

PLANTS AFFECTED Wide range of edibles and ornamentals
SEASON All year
WATCH FOR Dull, limp foliage
ACTION NEEDED? Yes, to quickly identify the cause

Foliage will wilt through a lack of water. Where this results from dry soil at the roots, the plant will quickly recover after watering. If this does not help, or plants wilt unexpectedly, then it may be a symptom of damage caused by root pests such as vine weevil larvae, root aphids, and cabbage root fly, or root diseases, like *Phytophthora* root rot or club root. Stem diseases, such as verticillium wilt, have similar symptoms. Check roots and stems for signs of trouble and use the appropriate treatment.

Wilting of rhododendrons and azaleas may be a symptom of *Phytophthora*.

WINTER MOTH *OPEROPHTERA BRUMATA*

PLANTS AFFECTED Fruit trees and deciduous trees and shrubs
SEASON Spring
WATCH FOR Holes eaten in leaves
ACTION NEEDED? Sometimes, to protect fruit trees

In winter and early spring, wingless female moths crawl up tree trunks to lay eggs. Pale green caterpillars hatch in spring to feed on new foliage and blossom, leaving it peppered with holes. Fruit trees can be protected using grease bands or tree barrier glue (see p.34) applied in late autumn, which prevents egg laying. Protect the vital pollinators in your garden by using insecticide before blossom opens and choosing an organic spray with natural pyrethrin that won't persist on the plant.

Winter moth larvae damage plants like hornbeam, but are usually held in check by birds.

BACTERIAL LEAF SPOT *PSEUDOMONAS* SPECIES

PLANTS AFFECTED Many species
SEASON Year-round
WATCH FOR Brown patches surrounded by a yellow halo
ACTION NEEDED? Yes, to prevent spread

Circular or straight-edged areas of brown tissue on leaves, usually encircled by a yellow halo, signal the presence of this disease. In extreme cases the infection will kill leaves and may also affect other parts of the plant: check stems for lesions or cankers if the leaves are affected. Bacteria spread via water splashed on to foliage, so water carefully at the base of plants rather than over their leaves. Remove affected leaves or stems to prevent spread. No chemical controls are available.

Infected leaves, such as this delphinium, can become brittle and die.

BEAN CHOCOLATE SPOT *BOTRYTIS FABAE*

PLANTS AFFECTED Broad beans
SEASON Spring to autumn
WATCH FOR Circular, dark brown spots
ACTION NEEDED? Yes, to prevent spread

Leaves of broad bean plants infected by this fungus develop dark chocolate-brown spots, which may also be seen on flowers and stems. In damp conditions, the spots spread and join together, causing leaves to die and stems to collapse. The disease can be spread on infected seed or via airborne spores. To avoid problems, don't save seed from infected plants, ensure broad beans are grown on well-drained soil, and space them at least 23cm (9in) apart to allow good airflow between plants.

At the end of each season, remove and destroy infected broad bean plants.

BEAN HALO BLIGHT *PSEUDOMONAS* SPECIES

PLANTS AFFECTED Dwarf French and runner beans
SEASON Summer and autumn
WATCH FOR Small spots surrounded by a bright yellow halo
ACTION NEEDED? Yes, to prevent spread

Leaves of dwarf French and runner beans may become marked with small watery spots that darken and get surrounded by bright yellow halos. These symptoms signal infection by bean halo blight, which may develop further as yellowing between leaf veins and, eventually, leaf fall. Soft, watery patches sometimes also appear on stems and pods. Remove infected plants after harvest and don't save their seed. Grow resistant varieties as there are no chemical controls for bean halo blight.

Water plants at the base because this bacterial disease is spread by water splash.

BLOSSOM WILT *MONILINIA* SPECIES

PLANTS AFFECTED Apple, pear, and *Prunus* trees
SEASON Spring
WATCH FOR Brown, withered blossom and leaves
ACTION NEEDED? Yes, to prevent spread

This fungal infection, which usually enters the stem through blossom, causes leaves at the shoot tips to turn rusty brown and wilt; the blossom itself also shrivels and browns. Buff-coloured pustules may be visible on diseased areas, especially in wet weather, which promotes the spread of spores. The same fungus also causes brown rot on fruit, so ensure rotten fruits are removed from trees in autumn. Spread can be reduced by pruning out and burning infected shoots.

The withered foliage on the shoot tip of this apple tree is a symptom of blossom wilt.

BOX BLIGHT *CYLINDROCLADIUM BUXICOLA*

PLANTS AFFECTED Box (*Buxus* species), especially hedges and topiary
SEASON All year round
WATCH FOR Pale brown leaves with white fungal growth beneath
ACTION NEEDED? Yes, to prevent infection and spread

This fast-spreading fungal disease can ruin established box plants. Leaves turn pale brown and may develop white fungal growth on their undersides during wet weather. They then fall, leaving bare patches where stems are stained black. Box blight spreads fastest in humid conditions, through water splash, contaminated garden tools, and from plants brought into the garden. Avoid overhead watering, sterilize cutting tools between plants, and quarantine new plants for a month before planting. Pruning in dry conditions also reduces the risk of infection.

CONTROL On lightly damaged plants, prune affected stems back to healthy growth, tidy the fallen leaves from beneath and within plants, and scrape away the surface layer of soil to remove infection sources. Badly damaged plants are best removed and disposed of by burning or in domestic waste. The fungicides tebuconazole and trifloxystrobin can be used as part of a control strategy, but will not wholly eradicate the disease on their own.

Where box plants are badly affected, you may be better off planting alternatives such as yew and berberis.

CLEMATIS WILT *CALOPHOMA CLEMATIDINA*

PLANTS AFFECTED Clematis, especially large-flowered varieties
SEASON Spring to autumn
WATCH FOR Wilting leaves, stems, and buds
ACTION NEEDED? Yes, to control spread

This fungus causes clematis plants to wilt suddenly from their tips, though not all stems may be affected. Leaves show dark spots and shrivel as their stalks turn black and dark streaks are visible inside diseased stems when cut. No chemical controls are available, so cut diseased stems down to healthy growth, sterilize pruning tools, and choose resistant species, such as *C. montana* and *C. macropetala*. Plant new clematis deeply to give them the best chance to thrive.

Dark tips signal clematis wilt. New, healthy shoots will often arise from buds below soil level.

DOWNY MILDEW *PERONOSPORACEAE*

PLANTS AFFECTED A wide range of soft-leaved plants
SEASON Year-round
WATCH FOR Discolouration with fuzzy fungal growth beneath
ACTION NEEDED? Sometimes, to control spread

Infections begin as small yellow or grey-brown patches on the upper surfaces of leaves. Fuzzy grey or white growth develops on lower surfaces and can spread quickly to kill whole leaves. Damp conditions and poor air circulation favour these fungus-like pathogens. To control infections, cut off affected leaves, allow more space between plants, remove weeds, and avoid overhead watering. Rotate veg crops to avoid reinfection from resting spores. No chemical controls are available.

Lettuces affected by downy mildew turn brown as the leaves die.

FIREBLIGHT *ERWINIA AMYLOVORA*

PLANTS AFFECTED Apples, pears, and related trees and shrubs
SEASON Spring to autumn
WATCH FOR Dead brown leaves at shoot tips
ACTION NEEDED? Yes, to prevent spread

The shoot tips of affected plants turn dry and orange-brown, as if scorched by fire. Symptoms usually occur as the fireblight bacteria travel back along the stem after entering through open blossom (which also withers and dies). Stems develop areas of sunken bark and a rusty red colour on wood beneath the bark. White ooze may appear at infected sites. Cut back affected stems at least 30cm (12in) into healthy wood, where there is no discolouration under the bark, and sterilize tools after each cut.

Fireblight does not respond to any chemical treatments.

GREY MOULD *BOTRYTIS CINEREA*

PLANTS AFFECTED Many, especially in greenhouses
SEASON All year
WATCH FOR Grey fuzzy mould
ACTION NEEDED? Yes, to prevent spread

This common fungus usually infects leaves through damage that is already present, but can also trouble healthy plants in humid conditions. Affected areas turn soft and brown with a covering of grey, fuzzy fungal growth. Moist air in greenhouses and in dense plantings favours the spread of the fungus, so ventilate well and avoid overcrowding. Remove any infected, damaged, or dead leaves promptly, and clean and disinfect the greenhouse annually. No chemical controls are available.

The fuzzy growth of grey mould can spread very quickly given moist conditions.

FUNGAL LEAF SPOT *VARIOUS FUNGI*

PLANTS AFFECTED A wide range of plants
SEASON All year
WATCH FOR Rounded grey or brown spots
ACTION NEEDED? Sometimes, to prevent spread

Various fungi can cause leaf spots, especially during damp weather or if air circulation is poor. The circular brown or grey spots appear on foliage, which may also turn yellow; tiny black or dark brown fungal fruiting bodies may be visible on spots. Many leaf spots cause little harm, but can indicate poor drainage or disease elsewhere. Remove affected leaves and rake up fallen leaves to prevent spread. Fungicide treatment is rarely necessary, but tebuconazole may provide some control on ornamental plants.

Brown or grey spots can spread and join to create larger patches of dead tissue.

PEACH LEAF CURL *TAPHRINA DEFORMANS*

PLANTS AFFECTED Peaches, nectarines, apricots, and almonds
SEASON Spring
WATCH FOR Bright red swelling and distortion on leaves
ACTION NEEDED? Possibly, to protect fruit trees

Leaves infected with this fungus develop swollen patches as they emerge in spring. These turn red or purple, before producing a white powdery coating and falling. Trees usually respond with a second flush of healthy leaves. Prevent fungal spores overwintering in the bark by removing infected leaves before the white bloom forms. Infection only occurs in wet conditions, so keep trees fan-trained against walls or fences dry under a frame covered with plastic sheeting from November until May.

The vigour of trees can be affected if they suffer peach leaf curl year after year.

PESTALOTIOPSIS *PESTALOTIOPSIS* SPECIES

PLANTS AFFECTED Mainly conifers, including hedges
SEASON Spring and summer
WATCH FOR Shoots turning brown from tips
ACTION NEEDED? Yes, to improve growing conditions

Foliage affected by this fungus turns yellow then brown at the shoot tips; whole shoots may die back. Tiny black fruiting bodies appear on affected areas and black threads of spores may be visible in wet weather. The fungus rarely kills its host, but reduces its vigour and spoils its appearance. Cut out affected shoots to reduce its spread. The fungicides tebuconazole, tricloxystrobin, and triticonazole may provide some control, but they are not specifically labelled for use against pestalotiopsis.

Conifers weakened by drought, aphids, or poor pruning are prone to infection.

POTATO BLIGHT *PHYTOPHTHORA INFESTANS*

PLANTS AFFECTED Potatoes and tomatoes (see p.97)
SEASON Summer
WATCH FOR Brown, watery spots spreading from leaf margins
ACTION NEEDED? Yes, to prevent spread

Leaves of affected plants develop patches of brown, watery rot, followed by white fungal growth, and finally collapse. The infection will spread into stems and tubers. To avoid problems, grow early varieties or those with blight resistance, and rotate crops. Earth up plants to protect tubers. Cut down infected tops to ground level and dispose of them in domestic waste or by burning. The tubers below can be harvested two weeks later. No chemical controls are available.

Wet summers create ideal conditions for the fungus-like organism that causes blight.

POWDERY MILDEW VARIOUS FUNGI

PLANTS AFFECTED A wide range of plants
SEASON Spring to autumn
WATCH FOR White, powdery coating on leaves
ACTION NEEDED? Sometimes, where plant growth is affected

Powdery mildews are fungal infections that cause an easily recognizable coating of dusty, white growth on the upper surface of the leaves of many edible and ornamental plants. Infection can also cause yellowing and distortion of foliage and may spread to other parts of the plant. Stressed plants are more susceptible to infection, particularly where they repeatedly experience drought. Plants often shrug off isolated patches of powdery mildew and even severe cases may not pose a problem on edible plants at the end of their growing season, so control measures are not always necessary. Grow resistant varieties where available.

CONTROL Remove infected leaves promptly. Take action to improve growing conditions and make plants less vulnerable. Water thoroughly and mulch around plants to retain moisture. Improve airflow by pruning, weeding, reducing the density of planting, and increasing the ventilation in greenhouses. Rake up fallen leaves in autumn to prevent reinfection the following year.

Each species of powdery mildew affects a narrow range of host plants, so cross-infection between plant groups is unlikely.

Effective fungicides are available to treat powdery mildew. Products containing both tebuconazole and trifloxystrobin are approved for edible crops and can also be used on ornamentals along with sprays containing triticonazole. Plant invigorators containing surfactants help control powdery mildew, and some gardeners spray plants with diluted milk.

RAMORUM DIEBACK *PHYTOPHTHORA* SPECIES

PLANTS AFFECTED Rhododendrons, viburnums, holm oak, and other shrubs and trees
SEASON All year
WATCH FOR Spreading brown lesions on leaves
ACTION NEEDED? Yes, it is a notifiable disease in the UK

Also known as sudden oak death, this disease is caused by two fungus-like pathogens from the genus *Phytophthora*. It has killed many native oaks in the USA and Japanese larches in the UK. Infected shrubs develop brown patches at leaf tips or margins which spread in a V-shape along the central leaf vein. Stems wilt and die back. The disease spreads readily in wet weather through contaminated soil, compost, or water.

Leaf tissue death is clearly visible on this rhododendron. Outbreaks of this disease should be reported to your local plant health authorities.

ROSE BLACK SPOT *DIPLOCARPON ROSAE*

PLANTS AFFECTED Roses
SEASON Spring to autumn
WATCH FOR Black or dark purple leaf spots
ACTION NEEDED? Yes, to prevent spread

This fungus causes ragged-edged black or dark purple spots to form on rose leaves, sometimes spreading over the upper surface. Black marks can appear on stems. Some black spot can be tolerated where roses are grown in mixed borders. Reduce infection levels by raking up and destroying fallen leaves to remove fungal spores, and pruning out marked stems before spring growth. Repeated application of triticonazole, tebuconazole, or trifloxystrobin can help to control the damage.

To prevent damage, choose varieties or older species roses with resistance where possible.

RUSTS VARIOUS FUNGI

PLANTS AFFECTED A wide range, from trees to vegetable crops
SEASON Spring to autumn
WATCH FOR Powdery orange, brown, or yellow pustules
ACTION NEEDED? Sometimes, to control damage

Rusts are common fungal diseases that affect many garden plants. Early symptoms include pale leaf spots that develop into small, raised orange, brown, yellow, white, or black pustules with a powdery appearance, usually on lower leaf surfaces. Pustules can be numerous, sometimes causing leaves to yellow and fall, thereby reducing the plant's vigour and occasionally killing it. Grow resistant varieties where available, and avoid overcrowding and overhead watering, which produce the moist conditions that promote infection.

CONTROL Many plants tolerate rust well without any intervention, but you should take steps to limit its spread and reduce reinfection. Weed borders regularly, because some weeds can harbour rust diseases that will spread to garden plants. Remove infected leaves promptly to slow the spread, unless this will result in a damaging loss of foliage. Tidy up all fallen leaves and diseased plant material at the end of each growing season to help reduce overwintering spores and place

There are thousands of species of rust fungi; most tend to infect mature plants, such as this broad bean.

them in green or domestic waste – spores will survive home composting. No fungicides are available to treat rusts on edible crops, but ornamental plants can be treated with tebucanozole, trifloxystrobin, or triticonazole.

SHOTHOLE VARIOUS

PLANTS AFFECTED Many plants
SEASON Summer and autumn
WATCH FOR Numerous small holes with brown edges
ACTION NEEDED? Yes, if only to identify the cause

The symptoms of this disease are small brown or brown-grey spots on the leaves; as the tissue within the spots dies, it falls away, leaving circular holes that are often brown at the edges. The disease affects a wide range of plants and can be caused by a variety of bacteria and fungi. It often has little effect on growth and requires no treatment, but it may be an indication of a more serious infection elsewhere, so check for other symptoms, particularly of bacterial canker (see p.134).

Cherry trees, both edible and ornamental, are susceptible to shothole.

SILVER LEAF CHONDROSTEREUM PURPUREUM

PLANTS AFFECTED Rhododendrons and *Prunus* trees
SEASON Summer to autumn
WATCH FOR Silvered foliage and dieback
ACTION NEEDED? Yes, to prevent infection and spread

Leaves of plants affected by this fungus develop a silvery sheen; branches die back and show brown staining at their centre when cut. Infection will spread through, and eventually kill, the plant. Dark fruiting bodies appear on dead wood and produce spores that are infectious between autumn and winter. Pruning susceptible plants in summer helps to avoid infection. Remove infected branches promptly, cutting back at least 15cm (6in) into healthy wood. No chemical controls are available.

A silver sheen is clearly visible on plums, but more subtle on other trees.

TOMATO BLIGHT PHYTOPHTHORA INFESTANS

PLANTS AFFECTED Tomatoes and potatoes (see p.95)
SEASON Summer
WATCH FOR Brown patches and shrivelling on leaves
ACTION NEEDED? Yes, to prevent spread

Blight is signalled by the appearance of brown, watery patches on foliage. These spread rapidly, causing leaves to collapse and dry out. Similar symptoms spread to stems and fruit, killing the plant and spoiling the crop. The same fungus-like pathogen causes tomato and potato blight. It is prevalent on outdoor plants in warm, wet weather; greenhouses provide protection, but don't make plants immune. Grow tomato varieties with blight resistance and pick off affected leaves promptly.

Dispose of infected leaves by burning or placing them in green waste bins.

TULIP FIRE BOTRYTIS TULIPAE

PLANTS AFFECTED Tulips
SEASON Winter to summer
WATCH FOR Buff patches, distortion, fuzzy fungal growth
ACTION NEEDED? Yes, to prevent spread

Emerging tulip leaves become distorted with buff patches of dead tissue. Plants may fail to flower and develop grey mould in damp conditions, which releases spores into the air. Black, pinhead-sized resting bodies form on dead tissue and overwinter on bulbs and in soil to infect plants the next year.

Plant healthy bulbs in early winter to reduce the chance of infection; don't replant tulips on diseased sites for at least three years. Remove infected foliage and bulbs promptly to prevent spread.

Withered tulip leaves appear burned, giving the disease its name.

TURF FUNGAL DISEASES

Fungal diseases can affect turf grasses and spoil your pristine green sward. Brown or bleached patches may result from dollar spot, red thread, snow mould, turf thatch fungal mycelium, and lawn rust. Problems can usually be kept at bay by good maintenance, which helps prevent the damp, humid conditions that encourage fungal growth and also increases the resilience of turf to wear and tear.

PLANTS AFFECTED Lawn grasses

SEASON Autumn and winter

WATCH FOR Brown or bleached patches, fungal growth, toadstools

ACTION NEEDED? Yes, to improve growing conditions

SYMPTOMS

Turf problems most often arise in autumn, especially in wet weather. Distinguishing between pathogens is difficult, because dollar spot, red thread, snow mould, turf thatch fungal mycelium, and lawn rust all cause pale yellow or brown patches of dead grass. Look closer for signs of distinctive fungal growth: thin pink-red strands for red thread; fuzzy, pale pink growth for snow mould; dense, pale fungal growth for turf thatch fungal mycelium; and orange or black pustules for lawn rust. There is no visible evidence of fungus for dollar spot. Large concentric rings of lush and yellowing grass are caused by fairy rings, which produce circles of toadstools in autumn. Other toadstools may appear randomly or over decaying roots.

PREVENTION

Sunny, free-draining conditions keep fungi at bay. Reduce shade by cutting back overhanging branches and mow regularly, collecting the clippings, to help reduce debris (thatch) which reduces airflow around plants. Scarify the lawn in spring and autumn to remove moss and thatch. Apply high-nitrogen lawn fertilizers during the growing season to boost growth, but stop in late summer to avoid disease-prone autumn growth. Alleviate poor drainage and compacted soil by spiking the lawn deeply with a fork at 15cm (6in) intervals.

CONTROL

You can treat red thread and snow mould by watering the fungicide trifloxystrobin on to your lawn. Remove toadstools as soon as they appear, and before they have a chance to release their spores. Fairy rings can only be removed by digging out and disposing of turf and soil from the area to a depth of 30cm (12in).

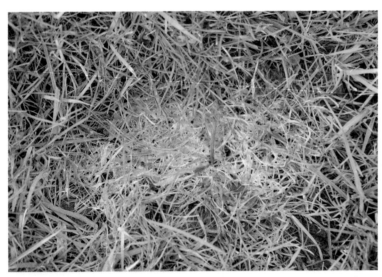

Turf snow mould (also called *Fusarium* patch) is often seen after snow cover, but also occurs in damp autumn weather. The slimy fungus creates brown patches on lawns.

Puffball mushrooms are common in lawns but do little damage.

VIRUSES

Viruses are tiny pathogens that infect a wide range of plants. Some have little or no visible impact, but others may result in loss of vigour, spoiling or reducing crops, or even killing plants. Take precautions to stop viruses entering your garden as little can be done once infection has occurred.

PLANTS AFFECTED A wide range

SEASON Mainly summer and autumn

WATCH FOR Foliage mottled, mosaic-patterned, streaked, deformed, or stunted

ACTION NEEDED? Yes, to prevent spread

SEE ALSO Page 107 for virus damage to flowers

SYMPTOMS

Characteristic symptoms of viral infection include yellow or pale green markings on leaves. You may see mottling, angular mosaic-like effects, or stripes. Patches of dead, brown tissue may also develop. Infected leaves are often small, distorted, or curled at the edges and new growth can have a stunted appearance. Plants often lose vigour, fail to flower, or have white streaks on petals (see p.107); crops from affected edible plants are often reduced. Common viruses include tomato viruses, strawberry viruses, and cucumber mosaic virus, which also affects lettuce, spinach, and ornamental perennials as well as cucurbits.

PREVENTION

Weeds may harbour viruses without showing symptoms. To prevent possible infection from these plants, you should weed your garden regularly. Viruses are immobile and rely on other organisms (vectors) to reach new plants. Vectors are often sap-sucking insects, including aphids and whiteflies, and eelworms which carry virus through the soil. Taking steps to control these pests will reduce the risk of infection, but won't remove it entirely. Humans are also significant virus vectors, spreading infection by handling, pruning, and propagating plants, so always wash your hands and disinfect tools before starting work in the garden.

Wherever possible, choose virus-resistant varieties or buy plants that have been certified as virus-free.

CONTROL

No treatments are available for infected plants. Ornamentals displaying mild symptoms may be worth keeping, but you should remove any that are growing poorly. Remove and dispose of all fruit and vegetable plants displaying symptoms of viral infection immediately. Avoid planting closely related plants together to hamper viral spread, and don't plant the same species in the same spot year after year.

Cucumber mosaic virus affects a wide range of garden plants.

A Michaelmas daisy shows how viral infection can distort growth, causing leaves to pucker, bend, or roll up.

BORON DEFICIENCY

PLANTS AFFECTED Mainly fruit and vegetable crops
SEASON Spring and summer
WATCH FOR Stunting or distortion with discolouration
ACTION NEEDED? Yes

Plants deficient in boron produce stunted, often distorted leaves with brown tips. Foliage of affected pear trees may die back at the shoot tips. Deficiency also causes splitting of stems (see p.140) and damage to root crops (see p.128). It arises on excessively limed soils, in dry conditions, or if boron has been washed out of light, sandy soils by rain. Apply borax or fritted trace elements (glass-based mineral supplements) to soil before planting, or use borax as a foliar spray for pear trees.

The foliage of affected plants may develop a yellowish or pinkish colouration.

DROUGHT

PLANTS AFFECTED Most plants, especially those in containers
SEASON Mainly spring and summer
WATCH FOR Wilting foliage and poor growth
ACTION NEEDED? Yes

The first sign of drought is usually dull or darkened foliage, followed by wilting. Longer dry periods will stunt growth, cause leaves to yellow, dry out, and fall and eventually lead to death. Prevent the soil drying out by watering thoroughly in the morning or evening (to minimize evaporation) during hot, dry weather; plants in containers may need daily watering. Incorporate organic matter into light soils to retain moisture and apply a thick spring mulch to prevent evaporation.

Plants in pots, in sandy or chalky soils, and young plants are most likely to be stressed.

FROST DAMAGE

PLANTS AFFECTED Many plants
SEASON Autumn to spring
WATCH FOR Brown or black foliage at shoot tips
ACTION NEEDED? Sometimes, to provide protection

Frost causes leaves at the tips of shoots to turn brown, black, or pale. Some foliage may also become darker green and look water-soaked, before turning black. The damage may be visible shortly after sub-zero temperatures, or it may take weeks to become apparent. Tender plants cannot tolerate frosts, while hardy plants are naturally adapted to survive icy winters; choose plants that are reliably hardy in your climate. Avoid planting susceptible plants in frost pockets or where exposure to early morning sun could increase damage. Grow more tender plants against a south-facing wall. Harden off half-hardy bedding and vegetables and wait for the risk of frost to pass before planting out.

CONTROL Cover vulnerable plants with horticultural fleece, either to extend the growing season of vegetable crops or to protect ornamental plants *in situ* for winter. Move tender plants growing in containers, or those that can be lifted from borders, under cover. Avoid using high-nitrogen fertilizers in late summer, as the soft growth they promote is easily damaged by frost. Delay cutting back more tender plants, like lavenders and penstemons, until spring. If frost damage occurs, cut out affected shoots once the risk of frost has passed.

Horticultural fleece allows light to reach plants, while providing a thermal blanket to protect against frost and pests.

IRON AND MANGANESE DEFICIENCY

PLANTS AFFECTED Mainly acid-loving plants
SEASON All year
WATCH FOR Yellowing between leaf veins
ACTION NEEDED? Yes

These deficiencies often appear together, affecting acid-loving plants such as rhododendrons, azaleas, and camellias planted on alkaline soil, where they are unable to absorb necessary nutrients. Symptoms include yellowing between the leaf veins, followed by browning of the leaf margins, usually on the youngest foliage first. Plant acid-lovers on appropriate soil or use ericaceous compost in pots. Mulch with acidic conifer bark and apply fertilizer formulated for acid-loving plants.

This camellia shows lime-induced chlorosis. This deficiency can be addressed using a treatment that contains sequestered iron and manganese.

MAGNESIUM DEFICIENCY

PLANTS AFFECTED Many plants
SEASON Spring to autumn
WATCH FOR Yellowing and other discolouration between leaf veins
ACTION NEEDED? Yes

The leaves of affected plants become yellow between veins and around margins, while areas surrounding veins remain green. The patches turn brown, red, or purple if the deficiency persists. The deficiency is common on light, sandy soils and where high-potash fertilizers (such as tomato feed) are regularly used. Treat affected plants during summer with a spray feed of Epsom salts applied in dull weather, and add Epsom salts to the soil to increase the availability of magnesium at the roots.

Older foliage shows symptoms of deficiency first, while young leaves remain unaffected.

NITROGEN DEFICIENCY

PLANTS AFFECTED Many plants
SEASON Spring and summer
WATCH FOR Yellow leaves, weak growth
ACTION NEEDED? Yes

Nitrogen deficiency is common because growing plants constantly deplete nitrogen reserves in the soil. Nitrogen is easily washed through light soils by heavy rain, and supplies in containers of compost are quickly exhausted.

Nitrogen is a nutrient that plays an important role in the production of new green leaves. In cases of deficiency, older leaves yellow first, often becoming tinged with pink or red, while new growth appears stunted and weak, and will start to yellow if the problem persists. Flowering, fruiting, and root growth can also be adversely affected.

CONTROL Nitrogen-rich fertilizers provide a short-term fix for ailing plants, and they should also be fed regularly to plants growing in containers. For a longer-term solution, add plenty of organic matter, such as well-rotted garden compost, to the soil when planting, and as a mulch every year thereafter. This will release nitrogen gradually as it breaks down, providing plants with a steady supply to fuel sturdy, healthy growth.

This yellowing strawberry has pink-tinged leaves – classic symptoms of nitrogen deficiency in a container plant.

Plants from the legume family (such as peas, beans, and clover) have root nodules that fix nitrogen from the atmosphere; leaving their roots in the soil after harvest, or growing them as green manures, is a good way to boost soil nitrogen levels.

OEDEMA

PLANTS AFFECTED Many plants
SEASON All year
WATCH FOR Raised corky growths on undersides of leaves
ACTION NEEDED? Yes, to prevent further damage

When plants take up more water than they can lose, leaf cells may bloat and rupture. Early symptoms of oedema are pale watery spots beneath leaves. These become raised and corky, and sometimes have a red-brown colour that is easily confused with rust disease. Wet conditions at the roots and poor air circulation are the main causes, so improve drainage, and grow plants further apart. Don't remove affected leaves or prune, as both will slow water loss and so worsen symptoms.

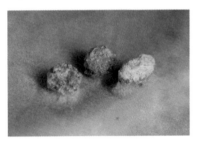

Crystal-like blisters on the undersides of leaves are a symptom of oedema.

PHOSPHORUS DEFICIENCY

PLANTS AFFECTED A wide range of plants
SEASON Spring to autumn
WATCH FOR Yellowing leaves and weak growth
ACTION NEEDED? Yes

Phosphorus is essential for the production of healthy roots and shoots. Deficient plants display slow, weak growth and their foliage looks dull, with a purple or blue-green tinge and yellow scorching at the leaf margins. This deficiency is rare, and is most likely on acidic or heavy clay soils and in areas where high rainfall leaches phosphorus from the soil. Treat by applying phosphate-rich fertilizers, such as superphosphate or bonemeal, to the soil where plants show symptoms.

This tomato leaf shows the purplish tinge associated with phosphorus deficiency. Leaves of affected plants may fall early.

POTASSIUM DEFICIENCY

PLANTS AFFECTED Many plants
SEASON Summer to autumn
WATCH FOR Yellowing leaves with brown margins
ACTION NEEDED? Yes

Potassium (potash) is a key nutrient for plant growth and promotes flowering, fruiting, and the ripening of growth for winter hardiness. Deficient plants develop yellow, sometimes purple-tinged leaves that turn brown and dry at the margins. Flowering and fruiting will also be poor. Deficiencies are most likely on light, sandy, or chalky soils, or in containers. Prevent and treat deficiencies by applying fertilizers high in potassium, such as tomato feed or sulphate of potash, or home-made comfrey tea.

Tips of leaves affected by potassium deficiency turn brown and curl up.

REVERSION

PLANTS AFFECTED Variegated shrubs and trees
SEASON Spring to summer
WATCH FOR Vigorous green foliage on variegated plants
ACTION NEEDED? Yes

Shoots with plain green foliage may appear among the (desired) yellow or cream patterned leaves of variegated shrubs and trees. Such "reverted" growth can occur in plants where the mutation that causes variegation is unstable. Pure green foliage has more photosynthetic pigments than variegated leaves, so it will grow vigorously and come to dominate the plant if left unchecked. Remove reverted shoots by pruning back to their base or into wood with variegated foliage.

Weigela is a plant often grown for its variegated foliage. Reversion spoils its appearance.

SCORCH

PLANTS AFFECTED Mainly plants with soft leaves
SEASON Spring to summer
WATCH FOR Pale, dry patches on leaves
ACTION NEEDED? Sometimes, to prevent further damage

Scorch marks on plants are caused by intense, hot direct sunlight, which damages tender foliage. Pale brown patches develop at the edges or in the centre of leaves. Damaged tissue usually dries out and becomes brittle, but other parts of the plant grow as normal. Water droplets on foliage, which focus the sun's rays onto the leaf surface, and hot greenhouse conditions will often lead to scorch. Shade greenhouses during summer, and water plants at their bases to avoid water droplets on leaves.

Scorch, as on this *Physalis*, is particularly common in greenhouse plants.

SUCKERS

PLANTS AFFECTED Roses and many trees, shrubs, and climbers
SEASON Spring to autumn
WATCH FOR Vigorous growth from roots
ACTION NEEDED? Yes

Suckers are new shoots sent up from plant roots. They are a natural way for plants to spread, but can spoil their look or cause a nuisance by growing up through lawns or borders. Where varieties of roses and trees have been grafted on to a rootstock, suckers have a different appearance to the existing plant and may be more vigorous. Remove unwanted suckers by digging down and removing them as close to the roots as possible; tear rather than cut to remove any buds at their base as well.

A sucker from a rose rootstock grows quickly and is very different in appearance.

WEEDKILLER DAMAGE

PLANTS AFFECTED Many plants
SEASON Spring to autumn
WATCH FOR Narrow, twisted foliage or scorched spots
ACTION NEEDED? Yes, to prevent contamination recurring

Weedkiller damage varies depending on the chemical involved. Leaves affected by glyphosate may be twisted, deformed, or stunted and show yellow or red discolouration, while contact weedkillers cause scorched brown spots. Damage may be on one side of a plant, where spray has drifted, or affect entire plants where contaminated sprayers or mulches have been used. Don't spray weedkiller in breezy conditions, use dedicated equipment, and rinse sprayers and cans thoroughly after use.

The effects of weedkiller cannot be reversed; your only option is to prune out badly damaged stems.

WIND DAMAGE

PLANTS AFFECTED Many plants
SEASON All year
WATCH FOR Battered, scorched, or brown leaves
ACTION NEEDED? Sometimes, where protection is possible

The soft new leaves of trees, shrubs, and climbers, along with young vegetable plants, are vulnerable to wind damage in spring and early summer. Foliage on the exposed side of a plant may turn dry and brown at the tips or edges. Larger soft leaves are often buffeted against stems or supports, causing tears or bruising that will turn brown. Wind has a drying effect on foliage which can exacerbate drought and frost damage. Plant sheltering trees and hedges to protect plants on windy sites.

The large leaves of runner beans are vulnerable to mechanical damage.

Flower pests can destroy a beautiful display, but avoid spraying flowers with insecticide to preserve pollinator populations.

FLOWERS AND BUDS

Flowers are at the heart of an ornamental display, and are no less vital when growing plants for food because pollination is necessary for fruits to set. Pests feed on flowers or cause distortion of shoot tips bearing buds and blooms, spoiling their appearance and affecting fruit development. Disease can spread into the rest of the plant via flowers, so watch for browning, withering, or mould. Cold or wet weather may damage blooms, while a lack of flowering is most often due to poor growing conditions and nutrient deficiencies.

APHIDS APHIDOIDEA

PLANTS AFFECTED Many trees, shrubs, perennials, and bulbs
SEASON Spring and summer
WATCH FOR Clusters of small insects
ACTION NEEDED? Yes

Aphid infestations are most apparent on leaves (see p.79) but can also affect bud development and cause flower deformities. Check plants regularly and squash clusters between your thumb and forefinger. Many beneficial creatures prey on aphids (see pp.28–29) but take a while to control the pests. Don't spray plants in flower with pesticide because of the risk to beneficial insects. Where essential, serious infestations can be sprayed while buds are green and tightly closed.

Roses can tolerate aphids to a degree, but act regularly to prevent their populations from building up.

APPLE SUCKERS PSYLLA MALI

PLANTS AFFECTED Apple trees
SEASON Spring
WATCH FOR Small insects present; blossom dying off
ACTION NEEDED? Yes, for heavy infestations

Tiny, pale green nymphs followed by aphid-like insects are visible in apple blossom. Insect-eating birds like blue tits may help to reduce numbers in spring, and a winter wash containing plant oils may help to remove overwintering eggs. If further treatment is required, spray with products approved for use on fruit trees, containing the synthetic pyrethroids deltamethrin or lambda-cyhalothrin, or the neonicotinoid acetamiprid. Apply while buds are green – never after flowers have opened.

Apple suckers draw sap from soft new growth. This can cause blossom to die back and poor fruit set where a tree is infested.

BIRDS VARIOUS SPECIES

PLANTS AFFECTED Fruit trees, fruit bushes, early spring flowers
SEASON Winter and early spring
WATCH FOR Missing buds or shredded blooms
ACTION NEEDED? Possibly, for crop plants

Most bird damage to flowers and buds occurs during winter and early spring when other food sources are limited. Species such as bullfinches strip dormant flower buds from fruit trees and bushes, reducing or destroying the crop. Other birds, including sparrows, peck at the spring flowers of ornamental plants, such as primulas and crocuses, spoiling the display but usually leaving plants intact. Secure netting over vulnerable plants; grow fruit bushes in a fruit cage.

Many species of birds, including finches, will feed on buds in spring.

CAPSID BUGS MIRIDAE

PLANTS AFFECTED Many shrubs and perennials
SEASON Late spring and summer
WATCH FOR Deformed shoot tips and aborted flower buds
ACTION NEEDED? Where damage is severe

These green insects are around 6mm (¼in) long and feed on the shoot tips and flower buds of many plants, including fuchsias, Caryopteris, Phygelius, and dahlias. Affected shoots may fail to develop flower buds, while damaged buds open unevenly. Light damage can be tolerated or pruned out. Where the problem is severe, spray at the first signs of damage with organic treatments containing natural pyrethrum or plant oils, which are less likely to damage beneficial insects than synthetic pyrethroids.

Bugs may be visible on plants or indicated by distorted leaves and leaves peppered with holes.

EARWIGS *FORFICULA AURICULARIA*

PLANTS AFFECTED Dahlias, chrysanthemums, and clematis
SEASON Late spring to autumn
WATCH FOR Ragged holes in petals and young leaves
ACTION NEEDED? Sometimes, to control numbers

Adult earwigs are about 16mm (⅔in) long, and have a pair of distinctive pincers at their rear. Active at night, they eat holes in the petals of young flowers. As the flowers grow, the holes expand into ragged tears. Earwigs also eat aphids, so tolerate their presence where possible. Earwigs can be collected on torchlit patrols or from inside straw-stuffed flowerpots placed upside-down on top of canes at flower height, where they will hide during the day. See also p.82.

Earwigs appear when plants are in flower, so do not use chemical sprays to control them.

SLUGS AND SNAILS GASTROPODA

PLANTS AFFECTED Many plants
SEASON Mainly spring and summer
WATCH FOR Damaged petals, slimy trails
ACTION NEEDED? Sometimes, to protect a floral display

Usually active at night and in damp conditions, slugs and snails will crawl up stems to feed on the soft petals of tulips, daffodils, and the flowers of bedding and many other plants. Their rasping teeth leave ragged holes, or flowers may be entirely eaten. Leaves may be damaged or are sometimes ignored. Pick the pests by hand after dark, and use biological controls and traps to reduce numbers. Scatter slug pellets containing ferric phosphate thinly around plants. See also pp.88–89.

Slugs will sometimes eat flowers in preference to foliage, ruining displays.

GALL MIDGES CECIDOMYIIDAE

PLANTS AFFECTED *Agapanthus*, *Aquilegia*, and *Hemerocallis*
SEASON Late spring and summer
WATCH FOR Distorted buds that fail to open and turn brown
ACTION NEEDED? Yes, to control spread

The activity of tiny flies called gall midges causes developing buds of daylilies (*Hemerocallis*) to become swollen with crinkled edges, and distorts the buds of columbines (*Aquilegia*) and African lilies (*Agapanthus*). Breaking the buds open reveals small white or yellow maggots which have hatched from eggs laid by the gall midges in late spring or early summer. Once maggots have fed on the bud they drop into the soil to pupate and overwinter, before emerging the following spring as the next generation of adult flies.

CONTROL Pick off and destroy all affected flower buds. Prompt action will prevent maggots from reaching the soil. Late-flowering daylily varieties produce buds after egg-laying has finished and are less likely to be affected. It is difficult to prevent female flies laying eggs, and the grubs are protected from chemical controls inside buds. Systemic insecticides containing the neonicotinoid acetamiprid may reduce damage, but must not be used while the plants are in flower.

This daylily bud shows the symptoms of gall midge infestation. Affected buds will fail to open and turn brown, while others on the same plant may develop normally.

BLOSSOM WILT _MONILINIA LAXA_

PLANTS AFFECTED Apple, pear, and _Prunus_ trees
SEASON Spring
WATCH FOR Brown, withered blossom
ACTION NEEDED? Yes, to prevent spread

Spring blossom infected with this fungus withers and turns brown, but remains attached to the shoot tip, rather than dropping off. Infection may move along the stem, causing foliage to wilt and turn rusty brown. The fungus is spread by wind-blown spores that cause infection during damp weather. Prune out and burn infected shoots and remove fruits with brown rot in autumn, as they can also be a source of infection. Choose resistant tree varieties where available.

Shrivelled, brown blossoms are a classic symptom of this fungal disease.

PEONY WILT _BOTRYTIS PAEONIAE_

PLANTS AFFECTED Herbaceous and tree peonies
SEASON Spring and summer
WATCH FOR Flower stems wilt and buds turn brown
ACTION NEEDED? Yes, to prevent spread

This fungal disease causes buds to turn brown and fail to open; the stems just below turn brown and collapse, causing the visible wilt. Affected areas develop fuzzy grey mould in damp conditions. Remove infected stems immediately by cutting back into healthy growth and burning the waste or sending it to landfill. This reduces the release of infectious spores and prevents black resting bodies (sclerotia) being produced to overwinter in the soil, from where they can reinfect plants the next spring.

Peony wilt affects buds, but wilt and browning may also be seen on leaves and stems.

GREY MOULD _BOTRYTIS CINEREA_

PLANTS AFFECTED A wide range of species
SEASON All year
WATCH FOR Small pale spots, brown patches with fuzzy growth
ACTION NEEDED? Yes, to prevent spread

Infection by spores of this fungus causes pale spots, sometimes edged with dark rings, to appear on petals. As the fungus spreads, petals and other flower parts rapidly turn brown, die back, and become covered with pale grey growth. Infection can quickly advance down stems and into the rest of the plant, so remove affected flowers promptly. Deadhead plants regularly as fading and damaged flowers are easily infected. Avoid overhead watering because spores are spread by water splash.

These campanula flowers show signs of damage by grey mould.

VIRUSES _VARIOUS_

PLANTS AFFECTED Many plants
SEASON Spring and summer
WATCH FOR White or paler streaks on petals
ACTION NEEDED? Yes, to prevent spread

The petals of various flowers can develop white or pale streaks. Such "colour breaks" caused by viral infection can sometimes be an attractive feature, but the number and size of flowers may be reduced and blooms may be distorted. No treatments are available for plants infected with viruses, so remove and destroy any that show symptoms to prevent spread by sap-sucking insects and other vectors. _See p.99 for further symptoms and information on controlling the spread of viruses._

Colour breaks in tulips may result in smaller flowers held on shorter stems.

BOLTING

PLANTS AFFECTED Leafy vegetables, onions, and root crops
SEASON Summer
WATCH FOR Early flowering
ACTION NEEDED? No

Tall flowering stems may appear prematurely before a useful crop of leaves has developed. Such bolting (or "running to seed") is usually caused by exposure to cold, changes in day length, or dry conditions, and will also occur at the end of an annual crop's growing season. To prevent bolting, delay spring sowings outdoors or make early sowings under cover in modules instead. Grow bolt-resistant varieties where available and improve growing conditions by adding organic matter every year.

Regular watering in hot summers will make plants like rocket less prone to bolting.

DROUGHT DAMAGE

PLANTS AFFECTED Many plants
SEASON Mainly spring and summer
WATCH FOR Wilting, browning, and bud drop
ACTION NEEDED? Yes, to prevent damage

Dry soil stops plants taking up adequate moisture for growth and development. Flowers, leaves, and stems wilt, but recover quickly from brief drought if water becomes available. Prolonged drought causes petals to turn brown and die, buds to dry or drop, and may prevent flower buds from forming. Light sandy soils dry out fastest. Help retain moisture by incorporating organic matter like garden compost every year, mulching in spring, and watering during dry weather.

These petunia flowers have been damaged beyond recovery by drought conditions.

FASCIATION

PLANTS AFFECTED Many plants, but a rare occurrence in most
SEASON All year
WATCH FOR Wide, flattened, distorted flowering stems
ACTION NEEDED? No

Fasciation is a disorder in which flowering shoots become flattened or curved and the flowers may emerge erratically along the stem. Daisy-like flowers appear elongated, with a fold along their central disc. The disorder is harmless and its appearance may even be an interesting curiosity. Fasciation may be caused by a mutation, a pest, cold weather, viral infection, or mechanical damage to the developing shoot. It does not usually recur, but plants such as *Primula* and *Veronicastrum* are particularly prone.

Flattened *Veronicastrum* shoots can be pruned out if necessary.

FROST DAMAGE

PLANTS AFFECTED Many plants
SEASON Autumn to spring
WATCH FOR Flowers and buds turning brown
ACTION NEEDED? Sometimes, to protect blossom

Frosty nights in autumn or spring may cause buds and open flowers to turn brown. Typically, only a proportion of blooms are damaged, because they are more exposed than others, or at a more delicate stage of development. Affected flowers on fruit trees and bushes may fail to set fruit. Site plants carefully, avoiding frost pockets and east-facing positions for those that flower in early spring. Cover plants with fleece overnight to provide protection if a damaging frost is forecast.

Damage may be worse if the morning sun causes rapid thawing after a frosty night.

POTASSIUM DEFICIENCY

PLANTS AFFECTED Many plants, especially those in containers
SEASON Summer and autumn
WATCH FOR Poor flowering
ACTION NEEDED? Yes, to rectify deficiency

Plants deficient in this essential nutrient produce few flowers, and those that appear tend to be small. Changes in foliage provide the best warning of deficiency: leaves yellow and develop purple tints and brown margins. Be vigilant if growing on light, sandy or chalky soils or in containers where the supply in the compost has been depleted. Deficiencies can be addressed by applying fertilizers high in potassium, such as liquid tomato feed or sulphate of potash.

Leaf yellowing provides an early warning of this deficiency. Applying fertilizer will encourage the formation of healthy flowers.

PROLIFERATION

PLANTS AFFECTED Many plants, including roses
SEASON Spring and summer
WATCH FOR Extra buds within flowers
ACTION NEEDED? No

This growth disorder may be caused by a mutation, pest activity, or cold weather. Numerous buds are seen growing from the centre of a flower, in place of pollen-bearing stamens. These buds may remain closed or open into full blooms; occasionally, extra leaves and stems are also present. It is unusual for all flowers on a plant to be affected by this phenomenon, but if all the blooms on a plant suffer every year, consider removing and replacing the plant.

Uncontrolled cell division produces new buds in the centre of a flower.

ROSE BALLING

PLANTS AFFECTED Roses
SEASON Summer
WATCH FOR Buds turning pale brown and failing to open
ACTION NEEDED? Yes, to prevent disease

This condition occurs when petals are wetted by rain then scorched by sunlight, causing them to die while leaving the bud beneath unharmed. The outer petals turn pale brown and dry, trapping the inner petals in a papery husk and preventing the flower from opening. Other buds on the same plant often open normally. Little can be done to prevent rose balling, but avoid overhead watering and prune out buds with symptoms to stop any secondary infections.

Affected flowers can become infected with grey mould in damp weather.

WIND DAMAGE

PLANTS AFFECTED Many plants
SEASON Mainly spring
WATCH FOR Dry brown marks on petals
ACTION NEEDED? Yes, to prevent disease

Damage often occurs on one side of the plant, where the flowers are exposed to the wind. It is most common in spring, when cold winds whip around delicate blooms held high on trees and shrubs. Nothing can be done once damage has occurred. Remove damaged flowers to prevent infection with secondary fungal diseases in damp weather. Provide shelter in windy gardens by planting hedges as windbreaks and locating spring-flowering plants in sheltered areas.

Open flowers that have been damaged by the wind become discoloured at the edges, usually turning brown and dry.

Ripe strawberries are a juicy target for a variety of garden pests.

FRUITS AND PODS

Harvesting delicious homegrown fruit and podded vegetables is a highlight of summer and early autumn. Plenty of pests are keen to feast on this bounty too, and given a chance they can severely impair the development of your crop. The thin skins and soft flesh of fruit can also be prone to fungal diseases, especially when already damaged by pests. Drought adversely affects fruit and pod development and, along with poor pollination and frost damage, is a common cause of fruit failing to set.

APPLE CAPSID *PLESIOCORIS RUGICOLLIS*

PLANTS AFFECTED Apples and pears
SEASON Late spring to summer
WATCH FOR Pale brown bumps on fruit skin
ACTION NEEDED? Rarely

Apples affected by these green, sap-sucking insects develop pale brown, slightly scabby, raised blemishes on their skin, though the flesh below remains undamaged and the fruit is still good to eat. Leaves at shoot tips may also be peppered with small holes. As damage is usually light and cosmetic, apple capsids can be tolerated. Encourage beneficial wildlife into the garden to control numbers; birds will feed on eggs overwintering in the bark of fruit trees.

Capsids cause marks and bumps on apples because they feed on developing fruits in early summer.

APPLE SAWFLY *HOPLOCAMPA TESTUDINEA*

PLANTS AFFECTED Apples
SEASON Spring and summer
WATCH FOR Black holes in fruitlets, scars on mature fruit
ACTION NEEDED? Yes, to control numbers

This pest makes holes in developing fruit, in which black excrement is visible. Mature fruits are marked with curving, yellow-brown scars. Apple sawfly larvae hatch from eggs laid on blossom, feed on fruitlets, then bore out to pupate in the soil. Pick off any affected fruit to stop the larvae reaching the soil. For serious attacks, use sprays approved for apple trees containing natural pyrethrins or synthetic pyrethroids. Spray within seven days of petal fall, but never on open blossom.

Mature apples showing scarring may look unappealing but remain edible.

BIRDS *VARIOUS*

PLANTS AFFECTED All soft fruit and tree fruit
SEASON Summer and autumn
WATCH FOR Holes pecked in fruit, berries missing
ACTION NEEDED? Yes, to prevent loss of crop

Birds, most notably wood pigeons and blackbirds, love soft fruit. As they pierce the skin and gouge into currants, berries, and tree fruits, they create entry points for other pests and fungal diseases.

Well before ripening, protect strawberries and bush and cane fruits within a fruit cage, or surround the plants with sturdy netting supported on a frame. Secure the netting carefully to stop birds becoming tangled in loose sections. Harvest tree fruit as soon as it is ripe to avoid damage.

This fig shows the deep gouges that feeding birds can make in fruit.

CODLING MOTH *CYDIA POMONELLA*

PLANTS AFFECTED Apples and sometimes pears
SEASON Summer to autumn
WATCH FOR Caterpillar at fruit core, dark hole in fruit
ACTION NEEDED? Yes, where damage is serious

A dark hole in the skin of the maturing fruits (which may ripen and fall early) is a sign of codling moth. When cut open, the cores look brown and tunnels made by the small white moth caterpillars may be visible in the flesh. Try hanging up pheromone traps in early May to catch male moths; this may prevent mating and so reduce caterpillar numbers. You can also spray in summer with natural pyrethrins and synthetic pyrethroids; they will only work if applied before the caterpillars enter the fruit.

Caterpillars feed at the core, then tunnel out to overwinter under bark or in leaf litter.

MICE *MUS* SPECIES

PLANTS AFFECTED Beans, peas, strawberries, and sweetcorn
SEASON All year
WATCH FOR Crops nibbled, stems of unripe strawberries cut
ACTION NEEDED? Sometimes, if damage is serious

Mice can cause considerable damage, especially if there are few predators to control numbers. They will raid pea and bean pods, especially if they are left on the plant to dry, and eat sweetcorn kernels from inside their husk. Strawberries are another favourite, and stored apples and pears can be partially eaten. Set mouse traps baited with peanut butter near fruit stores or damaged crops. Place them under a cover of bricks or logs outdoors to avoid harming birds or pets.

Shredded and discarded pea pods are a sign of mouse damage.

RASPBERRY BEETLE *BYTURUS TOMENTOSUS*

PLANTS AFFECTED Raspberry, blackberry, and other cane fruits
SEASON Summer
WATCH FOR Brown patches and small grubs
ACTION NEEDED? Sometimes, if crops are badly affected

Adult raspberry beetles lay their eggs on flowers in late spring and early summer. Their small, creamy larvae cause grey-brown, dry patches to form at the top end of fruits, and are often seen among picked fruit. Some damage is tolerable, and can be kept at low levels by encouraging predatory wildlife. Autumn-fruiting raspberries flower later and are usually unaffected. Avoid using pesticides, because effective control would require plants to be sprayed while in flower.

The larvae of the raspberry beetle are often visible on ripening fruit.

PEA MOTH *CYDONIA NIGRICANA*

PLANTS AFFECTED Peas
SEASON Summer
WATCH FOR Small caterpillars inside harvested pods
ACTION NEEDED? Sometimes, to prevent egg-laying

Pea moths are attracted to flowering pea plants to lay eggs in summer. When the pods are harvested, moth larvae – small, cream, black-headed caterpillars – can be seen feeding inside, with their droppings next to damaged peas. Prevent problems by making early spring and summer sowings that will flower outside the egg-laying period. Cover summer-flowering plants with horticultural fleece or fine insect mesh to exclude the adult moths. Do not apply insecticides.

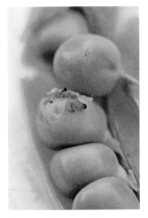

Pea moth caterpillars around 12mm (½in) long can be seen inside opened pods.

PLUM MOTH *GRAPHOLITA (CYDIA) FUNEBRANA*

PLANTS AFFECTED Plums, damsons, and gages
SEASON Summer and early autumn
WATCH FOR Caterpillars and their excrement inside ripe fruits
ACTION NEEDED? Sometimes, if crops are badly affected

These moths lay eggs on young shoots and leaves. The pale pink caterpillars that emerge tunnel into and feed on fruit. Affected fruits tend to ripen early, while later fruit may be unaffected. Use pheromone traps from mid-spring to catch male moths. This will reduce damage, but also indicate the time to spray (as only caterpillars outside fruits are killed). Spray small trees with approved natural pyrethrins or synthetic pyrethroids in early summer, and again three weeks later.

Caterpillars of the plum moth tend to feed close to the stone of the fruit.

PEAR MIDGE *CONTARINIA PYRIVORA*

PLANTS AFFECTED Pears
SEASON Late spring and early summer
WATCH FOR Blackened fruitlets falling from tree
ACTION NEEDED? Sometimes, if damage severe in previous years

These tiny midges can ruin developing pear fruitlets. They lay their eggs in pear blossom in spring. These hatch into cream-coloured maggots that feed inside fruitlets, which turn black from the base and drop from the tree in early summer. For minor attacks, simply pick blackened fruitlets off trees. Badly affected trees can be sprayed while blossom is in bud (but never open) using natural pyrethrums, fatty acids, plant oils, or synthetic pyrethroids labelled for use on pears.

Remove dead or fallen fruitlets to stop maggots from entering the soil.

STRAWBERRY SEED BEETLE *HARPALUS RUFIPES*

PLANTS AFFECTED Strawberries
SEASON Summer
WATCH FOR Small brown patches where seeds eaten
ACTION NEEDED? No

These glossy, black, nocturnal beetles eat seeds on developing strawberry fruits. The fruits continue to grow with brown patches where seeds are missing, but beetles may also eat into the fruit, encouraging other pests and secondary rots. The beetles also feed on weed seeds, so regular weeding will prevent them being attracted to strawberry beds. Empty jam jars sunk into the soil create pitfall traps that catch these beetles (along with other beneficial creatures, which can then be released).

The beetles are about 12mm (½in) long. No suitable chemical controls are available.

TOMATO MOTH *LACANOBIA OLERACEA*

PLANTS AFFECTED Tomatoes and some other greenhouse crops
SEASON Summer and autumn
WATCH FOR Green or brown caterpillars and leaf damage
ACTION NEEDED? Yes, to prevent serious damage

The pale green or brown caterpillars of this moth feed on the foliage and fruits of tomatoes grown outdoors and under cover. Newly hatched caterpillars feed together on the underside of leaves, but then strike out alone to eat holes in foliage and fruits. Watch out for eggs and young caterpillars and remove them promptly. Remove any overwintering cocoons and plant debris in your greenhouse. Several applications of a nematode biological control for caterpillars may be effective.

Holes made by caterpillars in tomato fruit may allow in secondary infections.

WASPS *APOCRITA SPECIES*

PLANTS AFFECTED Fruit trees, grapes, strawberries
SEASON Summer to autumn
WATCH FOR Hollows eaten into fruit
ACTION NEEDED? Sometimes, if damage is extensive

Wasps are familiar yellow and black striped insects, which are attracted to fruit as it ripens. They initiate damage on soft-skinned fruits, like plums and strawberries, gradually eating away hollows in the flesh, and will also enlarge holes left in tougher-skinned apples and pears by birds. Damage is usually limited and easy to tolerate, but where there are many, hang wasp traps in trees to tempt them away from fruit. Protect ripening trusses of plums and grapes in muslin bags.

Wasps are drawn to the sugary juice of apples that have been pecked by birds.

AMERICAN GOOSEBERRY MILDEW

PODOSPHAERA MORS-UVAE

PLANTS AFFECTED Gooseberries and blackcurrants
SEASON Summer
WATCH FOR Powdery white or pale brown growth on fruit skin
ACTION NEEDED? Yes, to prevent spread

Gooseberries affected by this fungus have a white powdery layer on their skin, which gradually turns pale brown. Fruits may be stunted, but remain edible after washing. Similar growth appears on the upper surface of leaves, especially at the shoot tips, which become twisted and stunted. Cut out mildewed shoots quickly and prune in winter to shorten shoot tips and produce an open structure of branches that allows air to circulate.

Nitrogen-rich fertilizers that promote soft growth should be avoided with gooseberries.

APPLE AND PEAR SCABS *VENTURIA* SPECIES

PLANTS AFFECTED Apples, pears, and related trees and shrubs
SEASON Spring to autumn
WATCH FOR Dark brown, scabby marks on fruit
ACTION NEEDED? Yes, to control spread

The appearance of small, dark brown, scabby patches on developing apples and pears, especially during wet growing seasons, signals infection by this fungus. Scabs often just cause superficial marks on the skin, but can result in distortion or cracked skin. Leaves also develop grey-brown spots and can fall early. Rake up affected leaves and fruitlets as they fall and prune out shoots with scabby bark where the fungus will overwinter. Choose resistant varieties, because no chemical controls are available.

Scabs can allow secondary infections to enter. Affected fruit may spoil when stored.

BROWN ROT *MONILINIA* SPECIES

PLANTS AFFECTED Apples, pears, plums, and other fruit trees
SEASON Summer and autumn
WATCH FOR Soft, spreading brown rot with buff pustules
ACTION NEEDED? Yes, to limit spread

This fungal disease manifests itself as orange-brown patches of rot that spread out from damage on fruits caused by birds, wasps, codling moth, diseases such as apple scab, and rubbing against branches. Rot may then spread to any fruit in contact with the affected fruit. Rings of pale pustules develop on rotten tissue as the fungus releases its spores. Affected fruits may fall from the tree early or remain attached to the branches, where they eventually dry out and shrivel into a "mummified" state. The same fungus causes blossom wilt in spring (see p.107).

CONTROL Reduce the number of overwintering spores on your trees by removing all infected fruits from the branches and the ground as soon as possible. The fruit will continue to rot and release spores, so dispose of it carefully either in a bin designated for green waste or by burying it at least 30cm (12in) below ground. Once branches are bare in early winter, prune out any mummified fruits with a short section of the

Greyish or cream pustules produce spores that can be transmitted by direct contact or by wind or insects.

shoot they are attached to. Reducing damage to fruit helps to prevent infection, so take action to control pests and other diseases where possible. No chemical controls are available.

GREY MOULD *BOTRYTIS CINEREA*

PLANTS AFFECTED Mainly soft fruits and fruiting vegetables
SEASON Summer to autumn
WATCH FOR Brown patches with fuzzy, grey growth
ACTION NEEDED? Yes, to limit spread

This common fungal disease infects fruit through wounds, but spores also enter open flowers, causing symptoms only once fruit ripens. Fruits turn soft and brown, and develop pale grey mould. Grey mould produces pale, circular discolouration on tomatoes, known as "ghost spots", and may cause stem dieback on gooseberries and cane fruits. Remove infected plant parts immediately to prevent spread and tidy up dead plant material that could harbour spores. Avoid overcrowding plants.

Cool, wet weather in late summer provides ideal conditions for grey mould.

POCKET PLUM *TAPHRINA PRUNI*

PLANTS AFFECTED Plums and damsons
SEASON Summer
WATCH FOR Flattened, elongated fruits without stones
ACTION NEEDED? Yes, to limit spread

This fungal disease causes young fruits to become pale and elongated, with no stone at their centre. White fungal growth covers the fruits, which then turn brown, shrivel, and fall from the tree or remain attached as dry "mummified" forms. Stems bearing these fruits may also be infected by the fungus and become distorted. Remove deformed fruits before they turn white and spores are released. Prune out any affected stems to prevent the fungus overwintering in the bark to infect future crops.

Fruits affected by pocket plum fail to ripen and are entirely inedible.

POWDERY MILDEW *VARIOUS FUNGI*

PLANTS AFFECTED Many soft-skinned fruits and peas
SEASON Summer
WATCH FOR White powdery patches
ACTION NEEDED? Yes, to control spread

Affected fruits and pods have pale powdery patches on their surface and are often small and distorted. Their skin may split, leaving them open to secondary infections. Plants stressed by drought and those growing in overcrowded or poorly ventilated greenhouses are particularly susceptible to this fungal infection. Ensure plants are well watered, with ample airflow. Remove affected fruit and leaves. Plant invigorators containing surfactants can help control powdery mildew on edible crops.

Fruits and sometimes leaves and flowers develop a white or grey powdery coat.

TOMATO BLIGHT *PHYTOPHTHORA INFESTANS*

PLANTS AFFECTED Tomatoes
SEASON Summer to autumn
WATCH FOR Brown areas beneath skin and rotting
ACTION NEEDED? Yes, to prevent spread

The fungus-like pathogen responsible for potato blight (see p.95) can also infect tomatoes, causing areas of fruit flesh to turn brown, collapse inwards, and rot. Brown patches also develop rapidly on foliage (see p.97). Blight needs moist conditions to infect plants, so is worse in warm, wet summers, and affects outdoor tomatoes more than those grown under cover. Remove any parts of the plant showing symptoms to slow spread. Grow varieties offering blight resistance outdoors.

Fast-spreading, watery brown rot is typical of tomato blight.

APPLE BITTER PIT

PLANTS AFFECTED Apples
SEASON Late summer and autumn
WATCH FOR Sunken brown spots on skin and in flesh
ACTION NEEDED? Yes, to reduce occurrence in following years

A lack of water during apple development can result in calcium deficiency. The skin of affected fruits becomes pitted with dark brown spots, and brown flecks appear throughout the flesh, which may develop a bitter taste. Mulch around trees, water regularly during summer, and feed with a balanced fertilizer. Apply calcium nitrate solution as a foliar spray during summer, but test a single branch first to avoid damaging sensitive varieties (such as Bramley's Seedling).

Distinctive spots on developing fruit are a result of calcium deficiency.

BITTER CUCUMBERS

PLANTS AFFECTED Greenhouse cucumbers
SEASON Summer
WATCH FOR Bitterness rendering fruits inedible
ACTION NEEDED? Yes, to prevent pollination

Greenhouse cucumbers sometimes taste extremely bitter. This taste only occurs in fruit that have developed from pollinated flowers. For this reason, modern varieties are "all-female"; however, older varieties bear both male and female flowers. In these varieties, just pick off the male flowers before they open: male flowers have a slender stem, where female flowers have a miniature "fruit". Note that outdoor ridge cucumbers need to be pollinated, so leave their male flowers intact.

Bitter cucumbers get their unappealing taste from the chemical cucurbitacin.

BLOSSOM END ROT

PLANTS AFFECTED Tomatoes, peppers, aubergines
SEASON Summer
WATCH FOR Dark patches at the base of fruits
ACTION NEEDED? Yes, to protect later fruit

Blossom end rot is caused by a lack of calcium, which is usually the result of dry conditions around the roots rather than a lack of the mineral in the soil or compost. Flattened black or brown patches can be seen at the base of developing fruits, where the flower once was. Patches vary in size from tiny spots to about 2.5cm (1in) across and may not occur on all fruits in a truss. The condition is most common on later trusses of fruit produced by plants grown in containers or growbags, which can easily dry out during hot summer weather. Plants growing in open soil are rarely affected.

CONTROL Pick off any affected fruits to prevent secondary infections setting in, and ensure that plants are watered consistently. Greenhouse crops and those in containers outdoors may need watering twice a day in hot summer weather. Install a drip irrigation system if you are unable to water regularly. Grow plants in larger containers if fruits have been badly affected in previous years.

Blossom end rot renders affected fruit inedible, but later trusses can be saved by watering more consistently.

FRUIT FAILS TO SET

PLANTS AFFECTED All fruit trees, soft fruit, fruiting and podded vegetables
SEASON Spring and summer
WATCH FOR Lack of flowering or young fruit forming
ACTION NEEDED? Yes, to identify cause

A variety of environmental factors, poor growing conditions, and nutrient deficiencies can all result in fruit and pods failing to set. It may take a while to work out the cause, but consider the following common problems. Spring frosts can badly damage open blossom, especially on early-flowering peaches, apricots, pears, and gooseberries. A windy site or enclosed greenhouse can prevent insects pollinating flowers, which is essential for the formation of many fruit and pod crops. Plants need good growing conditions to produce healthy crops, so a lack of water and nutrients in the soil can easily impair fruit set. Excessive or incorrect pruning can remove fruiting spurs or encourage vigorous leafy growth instead. Also watch for signs of pests and diseases thwarting fruit development.

CONTROL Protect bushes and wall-trained trees from late frosts with fleece, but remove covers during the day to allow pollination. Open doors and vents in greenhouses to

Frosts in spring can seriously damage early-flowering cherries, preventing fruit from setting.

encourage pollinating insects inside, mist flowering greenhouse crops to improve pollination, and employ hand pollination where practical. Ensure partner plants for cross-pollination grow nearby for fruit trees if required. Water plants regularly and apply mulch and fertilizer as appropriate. Follow pruning directions carefully to avoid removing fruiting spurs.

FRUIT SPLITTING

PLANTS AFFECTED Many fruits, especially tomatoes
SEASON Summer
WATCH FOR Splits in skin exposing flesh
ACTION NEEDED? Yes, to prevent the problem

Fruit skins will sometimes split for no obvious reason, leaving flesh inside exposed. Secondary infections can then set in, causing fruit to rot. Fruit splitting is usually the result of an erratic water supply, either caused by irregular watering or intermittent heavy rainfall. Regular watering and mulching to maintain consistent soil moisture will reduce the problem. Shading and ventilating greenhouses to avoid large temperature fluctuations also helps stop tomatoes splitting.

Thin-skinned fruits such as gooseberries, plums, and tomatoes split most frequently.

TOMATO GREENBACK

PLANTS AFFECTED Greenhouse tomatoes
SEASON Summer
WATCH FOR Green areas around top of ripe fruits
ACTION NEEDED? Yes, to prevent the problem

Greenhouse tomatoes may retain tough areas of green or yellow tissue at the top of the fruit. This condition is caused by exposure of the fruit to intense sunlight and high temperatures, accompanied by a shortage of potassium during ripening. Ventilate the greenhouse during the day to prevent extremes of temperature and apply shading to glass in summer. Don't remove leaves above fruit trusses until the end of the growing season. Feed regularly with a potassium-rich fertilizer.

The disorder is most common in heirloom and larger tomatoes.

Striped 'Chioggia Pink' beetroots show that roots can be ornamental and nutritious.

ROOTS, TUBERS, AND BULBS

Healthy roots absorb vital water and nutrients, and so are essential for strong growth. However, it is difficult to directly monitor their status below ground. Numerous soil-dwelling pests feed on roots, while diseases and disorders can cause serious damage where soil conditions are unsuitable, particularly if drainage is poor. Fleshy roots, tubers, and bulbs are important energy stores for many vegetable crops and ornamental plants, but also make attractive sources of food for pests.

BEAN SEED FLY *DELIA PLATURA*

PLANTS AFFECTED French and runner beans
SEASON Spring to summer
WATCH FOR Damage to seed leaves and growing tips of seedlings
ACTION NEEDED? Yes, to prevent damage

The white larvae of these flies feed on the roots of germinating bean seeds. The seedlings emerge with brown marks and pits in their seed leaves and stems, as well as damage to their growing tips, or may fail to emerge at all. Reduce problems by adding organic matter to soil in autumn rather than spring, sowing in warm soil so that seedlings grow rapidly, and covering sowings with insect mesh or fleece to prevent adult flies laying eggs in soil. Plants raised in pots or modules will avoid damage.

Brown, pitted seedlings are the classic result of bean seed fly activity.

CABBAGE ROOT FLY *DELIA RADICUM*

PLANTS AFFECTED Cabbages, kale, broccoli, and other brassicas including root crops and ornamental plants
SEASON Spring to autumn
WATCH FOR Slow growth, wilting, tunnels in root vegetables
ACTION NEEDED? Yes, to prevent damage

The maggots of these flies feed on roots, often eating them down to the stump. Members of the cabbage family, especially young transplants, are worst affected and may lack vigour, wilt, and die. Maggots also tunnel inside brassica root crops, such as turnips and swedes. Practise crop rotation and prevent flies laying eggs at the base of young plants by covering with insect mesh or fleece, or using collars round stems (see *p.34*).

The pale larvae are up to 8mm (⅓in) long. They feed on roots before tunnelling up into the main plant stem.

CARROT FLY *PSILA ROSAE*

PLANTS AFFECTED Carrots, parsnips, and parsley
SEASON Summer to autumn
WATCH FOR Brown tunnels in roots with small cream-coloured larvae inside
ACTION NEEDED? Yes, to prevent damage

The fleshy tap roots of affected carrots and related crops display long, brown, indented marks on their surfaces. When cut open, the roots are often criss-crossed by further tunnels, which may contain the small, cream-coloured carrot fly larvae responsible. The damage can render large proportions of affected crops inedible and also allow secondary rots to set in. Several generations of adult carrot flies lay eggs on the soil close to crops from late spring to autumn. Carrot flies overwinter in the soil as larvae or pupae.

CONTROL Protect crops in the carrot family by rotating them around the vegetable plot (see *p.33*), so that they are sown into soil where there are no overwintering larvae or pupae. Cover the plants with fine insect mesh or fleece to prevent adult flies gaining access to the soil to lay eggs. Adult carrot flies are low-flying, so 60cm- (2ft-) high barriers of fleece around carrots will greatly reduce the numbers of eggs laid.

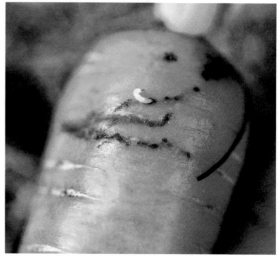

Biological controls which use nematodes can combat carrot fly larvae, but there are no available pesticides.

Sow carrot seeds thinly to reduce the need for thinning, because egg-laying adults are attracted by the scent of uprooted seedlings. Some carrot varieties are less susceptible, but not immune, to damage: try 'Resistafly' or 'Sytan'.

CUTWORMS VARIOUS MOTH LARVAE

PLANTS AFFECTED Mainly seedlings of annual plants
SEASON Spring to autumn
WATCH FOR Seedlings and young plants severed at soil level
ACTION NEEDED? Yes, to control damage

Fat, soil-dwelling, brown, green-brown, or cream moth caterpillars can cause damage by feeding on roots and stem bases of young plants, leaving severed stems and wilted leaves. Dig around damaged plants to find and remove the culprits. Keep soil well weeded and watered and encourage beneficial wildlife to control numbers. Grow veg under fleece or insect mesh to prevent adult moths from laying eggs. A nematode biological control can be watered onto affected areas.

Cutworms will feed on a great variety young seedlings or transplants

CHAFER GRUBS PHYLLOPERTHA HORTICOLA

PLANTS AFFECTED Lawns, herbaceous perennials, and vegetables
SEASON Spring to autumn
WATCH FOR Yellow patches in turf, wilting plants
ACTION NEEDED? Sometimes, to control lawn damage

These grubs – the larvae of chafer beetles – are most obvious when they affect turf, producing yellow patches where their feeding has damaged roots. They also feed on roots of herbaceous perennials and vegetables, but this damage is usually light. Foxes, badgers, and birds compound damage by ripping up lawns to feed on the grubs. Remove any larvae found in borders and reduce lawn infestations by watering on a nematode biological control.

These large, creamy, C-shaped beetle grubs have a brown head and three pairs of legs.

EELWORMS NEMATODA

PLANTS AFFECTED Potatoes, tomatoes, and numerous ornamental plants
SEASON Late summer and autumn
WATCH FOR Yellowing foliage, stunted growth, small cysts on roots
ACTION NEEDED? Yes, to prevent spread

Eelworms (nematodes) are 1mm- (¹⁄₂₄in-) long worms that feed within plant roots. The damage they do prevents the movement of water and nutrients around the plant, resulting in the yellowing of foliage, beginning with the oldest leaves. Plants may wilt and their growth becomes slow and stunted, and potato plants will produce poor crops. If affected plants are lifted carefully from the ground, tiny cyst-like growths will be visible on their roots, similar in appearance to the larger nitrogen-fixing nodules found on roots of healthy peas and beans. The cysts are formed by female eelworms and contain hundreds of eggs; the eggs of potato cyst eelworms can persist in the soil for many years, until stimulated to hatch by chemicals released by the roots of potato plants.

CONTROL No chemical or biological controls are available, so remove affected plants and their surrounding soil immediately and dispose of them in domestic waste or by burning.

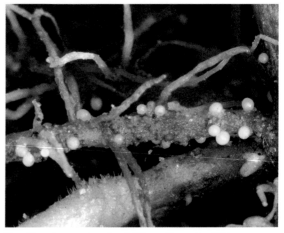

Small cysts are clearly visible on the roots of this infected potato plant.

Disinfect all tools and footwear afterwards to prevent spread. Practice the longest possible crop rotation to help stop pest populations building up in the soil. Grow resistant varieties of potato: these are typically resistant to only one nematode species and will allow populations of other species to persist, but are nevertheless a useful measure.

FUNGUS GNATS *ORFELIA* AND *BRADYSIA* SPECIES

PLANTS AFFECTED Seedlings under cover
SEASON Year-round
WATCH FOR Root damage, wilting, adult flies on compost surface
ACTION NEEDED? Yes, to control numbers

Adult fungus gnats are small brown flies that scurry over the surface of compost where plants are grown under cover. Their slim, pale larvae live in compost, feeding on decaying organic matter and sometimes soft seedlings or cuttings. Established plants are not harmed. Control numbers by removing dead plant material from pots and avoiding overwatering, Catch adult flies using sticky yellow traps. A biological control containing predatory mites is available.

Larvae of the fungus gnat are semi-transparent and about 6mm (¼in) in length. They thrive in wet compost.

LEATHERJACKETS *TIPULA* SPECIES

PLANTS AFFECTED Annuals, vegetables, and lawns
SEASON Spring and summer
WATCH FOR Young plants collapse; yellow patches in turf
ACTION NEEDED? Yes, where damage is serious

These grey-brown, legless grubs are the larvae of crane flies, which do not damage plants as adults. They eat the roots of seedlings, young plants, and lawn grasses, causing them to wilt, yellow, and often collapse. Further damage can also be caused to lawns by birds pecking at turf to feed on larvae in the soil. Birds and other wildlife will help to control numbers. Where damage cannot be tolerated, biological controls containing nematodes can be watered onto soil or lawns to kill larvae.

Leatherjacket damage is most often seen in summers that follow mild winters.

MICE *APODEMUS* SPECIES

PLANTS AFFECTED Vegetables, flowering bulbs
SEASON Year-round
WATCH FOR Shoots left on soil surface, nibbled root vegetables
ACTION NEEDED? Yes, if damage persists

Mice are great opportunists. While they love to eat sowings of nutritious pea, bean, and sweetcorn seeds (see p.112), they will also feed on any exposed surfaces of root crops left in the soil as the weather turns colder in autumn and on newly planted flower bulbs, such as crocuses, which they will unearth. Firm soil well after planting to make it harder for mice to excavate bulbs. Set mouse traps around vulnerable plantings, covered with bricks to protect pets and birds.

Mice will burrow through loose soil in search of starchy potato tubers.

NARCISSUS BULB FLY *MERODON EQUESTRIS*

PLANTS AFFECTED Mainly daffodils
SEASON Late summer to spring
WATCH FOR Weak growth or failure to emerge in spring
ACTION NEEDED? Yes, to prevent adults laying eggs

If your daffodils fail to grow in spring or only produce spindly leaves, check for the presence of narcissus bulb fly larvae. These pale brown grubs feed at the centre of each bulb and fill the void with brown droppings. Dispose of affected bulbs and prevent adult flies laying eggs in early summer by piling extra soil around the necks of healthy bulbs or covering them with insect mesh from late spring. Avoid growing daffodils in sheltered, sunny sites. No chemical controls are available.

Narcissus bulb fly larvae are found singly. They are about 15mm (⅔in) long.

ONION FLY *DELIA ANTIQUA*

PLANTS AFFECTED Onions, shallots, leeks, garlic
SEASON Summer
WATCH FOR Seedlings dying, yellowing leaves, damaged bulbs
ACTION NEEDED? Yes, to prevent eggs being laid

The small, white maggots of the onion fly feed on allium roots and tunnel into bulbs, making them susceptible to secondary infections. Seedlings may collapse and die, while the foliage of affected mature plants will wilt and turn yellow. Adult flies lay their eggs on or near plants from late spring until late summer, so protect crops under fleece or insect mesh during this period. Remove affected plants promptly, before larvae burrow into soil to pupate. Do not compost affected plants.

The pale larvae of the onion fly can be seen here feeding at the base of an onion bulb.

ROOT APHIDS *APHIDOIDEA*

PLANTS AFFECTED Lettuce, beans, and other veg and ornamentals
SEASON Mainly summer
WATCH FOR Plants wilting readily, slow growth
ACTION NEEDED? Sometimes, to control damage

If your plants lack vigour and wilt unexpectedly in warm weather, check their roots for these sap-sucking insects. Remove badly infested plants and practice crop rotation to stop reinfection by overwintered insects. Most chemical controls are ineffective because root aphids are hidden under the soil, but a soil drench containing the systemic neonicotinoid insecticide acetamiprid may be used to treat ornamental plants in containers.

Root aphids vary from pale yellow to dark brown and may secrete a white, waxy substance.

SLUGS *GASTROPODA*

PLANTS AFFECTED Potatoes, root vegetables, tulip bulbs
SEASON Summer and autumn
WATCH FOR Circular holes in tubers and roots
ACTION NEEDED? Sometimes, to control damage

Slugs will make circular holes in potato tubers, root veg, and tulip bulbs. Cutting in often reveals that the holes open out into cavities. Secondary infections often cause the vegetables or bulbs to rot in the soil or later in storage. The slugs responsible are soil-dwelling, so can't be controlled using traps or slug pellets. Grow varieties that are less prone to damage and lift crops as soon as possible. A nematode biological control applied to warm, moist soil will reduce slug numbers.

Slug damage is usually only noticed after potatoes have been lifted.

SQUIRRELS *SCIURUS* SPECIES

PLANTS AFFECTED Ornamental bulbs
SEASON Autumn to spring
WATCH FOR Signs of digging where bulbs planted
ACTION NEEDED? Yes, to prevent damage

Squirrels will excavate and eat dormant bulbs from borders and containers, often during autumn when they are newly planted. They will visit gardens repeatedly to remove whole plantings of crocuses and tulips once these are discovered. Nothing can be done to keep squirrels out of gardens. Protect bulbs after planting by covering the soil or container with well-secured chicken wire or other wire mesh, which can be removed in spring. Squirrels will bite through plastic netting or mesh.

Grey squirrels are widespread and can be a pest in urban and rural gardens.

SWIFT MOTH LARVAE HEPIALIDAE

PLANTS AFFECTED Herbaceous perennials and bulbs
SEASON Mainly spring to autumn
WATCH FOR Lack of vigour, damage to roots
ACTION NEEDED? Sometimes, to reduce damage

If your plants unexpectedly lose vigour, check their roots for long, slender, cream-coloured caterpillars with brown heads. These are swift moth larvae, which live in soil and feed on roots and stem bases. Remove any larvae you find while cultivating soil and encourage beneficial wildlife into the garden to help control numbers. Weed regularly, because adult moths prefer to lay eggs where soil is covered in vegetation. Treat soil with a nematode biological control for use against caterpillars.

Larvae are around 40mm (1½in) long. Their activity saps vigour from established plants, but rarely kills them.

WIREWORMS ELATERIDAE

PLANTS AFFECTED Root crops, seedlings, and many other plants
SEASON All year
WATCH FOR Tunnels in root veg, severed seedlings, root damage
ACTION NEEDED? Sometimes, to control damage

Ochre-coloured wireworms are the larvae of click beetles. They tunnel into potatoes and other root crops and may also sever seedlings at soil level and feed on roots. They usually occur where new beds are created from grassed areas and numbers naturally decrease over several years. Remove larvae during cultivation and reduce numbers by burying old potatoes next to developing root crops to attract larvae, then disposing of them. Nematode biological controls are also effective.

Wireworms are ochre to orange in colour, around 25mm (1in) long, and have three pairs of short legs.

VINE WEEVIL LARVAE OTIORHYNCHUS SULCATUS

PLANTS AFFECTED Many plants, particularly those in containers
SEASON Late summer to spring
WATCH FOR Poor growth and unexplained wilting
ACTION NEEDED? Yes, to limit damage

These beetle larvae feed below ground on a broad range of plants, but favour those with fleshy roots, including strawberries, *Primulas*, and *Heucheras*. Affected plants will show declining vigour, yellowing foliage, and sudden wilting, and often die. When lifted, it is easy to see that their root system has been eaten away. A search of the surrounding soil will most likely reveal small, plump, white grubs with curved bodies and brown heads. The dull black adult weevils are about 10mm (⅜in) long and have dirty yellow marks on their wing cases: these beetles are far less harmful than their larvae, but eat notches from the foliage of shrubs and herbaceous plants (see p.91).

CONTROL Remove and squash any grubs found in soil or compost. Encourage beneficial wildlife into the garden to prey on both adult beetles and larvae. Take measures to control numbers of adult beetles during spring and summer (see p.91). Biological controls containing nematodes can be watered onto

The larvae can be seen when raking through soil around damaged plants. They may affect bulbs as well as roots.

open soil or containers in late summer or early autumn to kill larvae before they cause damage. A soil drench containing the systemic neonicotinoid insecticide acetamiprid may be used to treat ornamental plants in containers.

CLUBROOT *PLASMODIOPHORA BRASSICAE*

PLANTS AFFECTED Vegetables from the cabbage family and related ornamentals and weeds
SEASON Summer to autumn
WATCH FOR Swollen, deformed roots and weak growth
ACTION NEEDED? Yes, to aid plant growth and prevent spread

The first visible symptom of clubroot is typically pale foliage, tinged pink or purple, that wilts readily. Growth is generally weak, vegetables produce a poor crop, and affected plants may die. These above-ground symptoms are the result of infection of the roots by a fungus-like pathogen, which causes them to become dramatically swollen and distorted, reducing their ability to take up water and nutrients. When affected roots disintegrate, they release thousands of spores into the surrounding soil. These are very easily spread on tools and boots, and can remain viable for up to 20 years, even when host plants are not present.

CONTROL Remove and burn infected plants immediately, before they have a chance to release their spores. Acidic and poorly drained soils favour clubroot, so add lime to soil in autumn to increase pH, and improve drainage by incorporating organic matter into the soil or by growing

Dig up affected plants and remove the entire root system to prevent spread. Never compost infected plants.

in raised beds. Choose vegetable varieties with resistance to the disease. Sow them in pots so they can develop a large, strong root system before planting them out. Take care not to introduce spores into your garden. The best way is to raise brassica plants from seed or to buy young plants from reputable sources. No chemical controls are available.

CROWN GALL *RHIZOBIUM RADIOBACTER*

PLANTS AFFECTED Many woody and herbaceous plants
SEASON All year
WATCH FOR Knobbly swellings on, or just above, roots
ACTION NEEDED? Yes, to prevent spread

This bacterial disease causes clusters of rounded swellings, known as galls, to develop on plant roots. The infection enters through wounds: growth is often unaffected at first, but as the galls disintegrate, they leave the damaged tissue open to secondary infections. Remove infected plants promptly to avoid bacteria spreading in the soil. Grass over the affected area for at least two years or grow resistant plants, such as potatoes, while the bacteria die off.

Galls can vary in size from about 1cm (⅖in) across to a diameter that engulfs the whole root system.

FOOT AND ROOT ROT *VARIOUS FUNGI*

PLANTS AFFECTED Soft-stemmed ornamental, fruit, and veg plants
SEASON Spring to autumn
WATCH FOR Dark brown, soft roots; wilting and yellowing foliage
ACTION NEEDED? Yes, to prevent spread

The roots of infected plants turn dark brown, soften, and rot away, causing plants to wilt, yellow, and die. The fungi responsible produce similar symptoms at the base of stems, often leading to plant collapse. Foot and root rots are common in young plants and greenhouse fruiting crops, such as tomatoes and cucumbers. Clean equipment thoroughly, use sterilized compost, and water young plants with tap water where possible as pathogens may be present in non-mains water.

This section through a broad bean plant shows rot affecting the root and stem.

MOULD ON BULBS VARIOUS FUNGI

PLANTS AFFECTED Many flowering bulbs
SEASON Spring to autumn
WATCH FOR Brown marks and grey or blue fungal growth
ACTION NEEDED? Yes, to prevent spread

Stored bulbs may develop red-brown patches, followed by blue-green fungal growth and sometimes rotting. To prevent these problems, store bulbs in a cool, dry place, with good ventilation.

If bulbs grow weakly or fail to emerge, tulip grey bulb rot may be the cause. Bulbs rot as a result of pale grey fungal growth on which large, black resting bodies may be visible. Remove affected plants and surrounding soil and don't replant bulbs in the area for at least five years. No chemical controls are available.

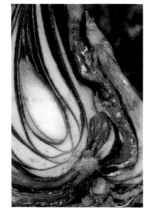

Blue mold can be seen on this lily bulb. It is caused by a species of *Penicillium*.

ONION WHITE ROT STROMATINIA CEPIVORA

PLANTS AFFECTED Onions, shallots, garlic, and leeks
SEASON Summer
WATCH FOR Rotten roots, white fungal growth at base of bulb
ACTION NEEDED? Yes, to prevent spread

This fungal disease causes rot in bulbs of the onion family. This becomes evident above ground as wilting, yellowing foliage. White fungal growth, with black, seed-like resting bodies can be seen on the bases of bulbs. Remove and burn infected plants and don't grow plants from the onion family in the same bed for at least eight years. No chemical control is available. Avoid introducing the fungus in contaminated soil on boots, tools, or young plants; buy onion sets and garlic bulbs from reputable suppliers.

White rot is visible on these shallots. The disease can remain in the soil for years.

PARSNIP CANKER ITERSONILIA PASTINACAE

PLANTS AFFECTED Parsnips
SEASON Autumn and winter
WATCH FOR Orange-brown marks around top of roots
ACTION NEEDED? Yes, to prevent infection

Orange-brown patches may develop on the pale skin of parsnips, most often around the top of the root. This damage can be cut away before eating, but badly affected roots are prone to secondary rots. Parsnip canker is a widespread fungal infection. It occurs when roots are injured and can follow damage by carrot fly larvae. Grow varieties with canker resistance, avoid damaging roots during weeding, protect parsnips from carrot fly with fleece or insect mesh, and improve soil drainage.

Canker spots may be small and smooth or large with a roughened surface.

PHYTOPHTHORA ROOT ROT PHYTOPHTHORA

PLANTS AFFECTED Shrubs and trees, including conifers
SEASON All year
WATCH FOR Blackened, rotting roots, foliage wilting and yellowing
ACTION NEEDED? Yes, to prevent spread

The fungus-like pathogen *Phytophthora* is a common cause of rot. Fine roots quickly rot away and larger roots blacken and are easily snapped. Foliage wilts and yellows, branches die back, and plants may be killed. It occurs in waterlogged, heavy soils, where its spores can persist for several years. Remove infected plants and the soil surrounding their roots. Improve drainage and replant with specimens not susceptible to infection. No chemical treatments are available for *Phytophthora*.

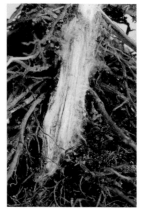

Dark brown discolouration can sometimes be seen beneath the bark at the stem base.

POTATO BLIGHT *PHYTOPHTHORA INFESTANS*

PLANTS AFFECTED Potatoes
SEASON Late summer and autumn
WATCH FOR Dark, sunken patches on skin, red-brown flesh, rotting
ACTION NEEDED? Yes, to prevent spread

Potato tubers affected by this fungus-like pathogen show slightly indented, spreading dark patches on their surface. The flesh beneath these patches becomes discoloured with red-brown markings. Secondary rots rapidly set in, causing the tubers to turn soft and slimy, collapse, and develop a foul smell. Phytophthora spores spread to potato tubers by washing down into the soil from infected foliage. Blight is therefore more prevalent in warm, wet summers, when conditions are best for infection and spread.

CONTROL Earth up potato plants deeply by pulling soil up into mounds around their stems to reduce the chance of spores from the leaves reaching the tubers. Grow early potato varieties, which are usually harvested before blight occurs, or choose varieties with blight resistance. Practice crop rotation to prevent carrying infection over into the next growing season. Cut down the top growth of infected plants to ground level promptly and place in green or domestic

Blight makes potatoes inedible. It is best to remove and destroy infected plants and tubers.

waste to prevent home compost becoming a source of infection. Lift the tubers two weeks later, once their skin has toughened, to reduce the chance of infection. No chemical controls are available.

POTATO COMMON SCAB *STREPTOMYCES SCABIEI*

PLANTS AFFECTED Potatoes, beetroot, swedes, turnips, radishes
SEASON Summer
WATCH FOR Dark, rough, raised patches on skin of tubers
ACTION NEEDED? Yes, to reduce symptoms

Affected potatoes develop dark brown, rough, scab-like patches on their surface, which sometimes form shallow cracks. Damage is usually superficial and the flesh remains good to eat. Common scab is caused by a bacterial infection that tends to be worse in dry and alkaline soil. Incorporate organic matter into the soil to help retain moisture, water plants regularly from about two weeks after they emerge, avoid liming the soil, and choose resistant potato varieties.

Common scab is most prevalent during dry summers.

POTATO GANGRENE *PHOMA FOVEATA*

PLANTS AFFECTED Potatoes in storage
SEASON Late summer to winter
WATCH FOR Sunken areas of skin, rotting tubers
ACTION NEEDED? Yes, to prevent spread

This slow-growing fungal disease affects stored potatoes. The tubers develop depressions on their surface, as if someone has pushed their thumb into the skin. The skin over these areas turns dry and wrinkled, and the cavities may be lined with white fungal growth. Avoid damaging tubers when harvesting and do not store damaged tubers. Inspect stores regularly and remove any potatoes that show symptoms to prevent infection spreading. Don't plant affected tubers.

Tubers will appear wet as they start to rot. Tiny black fruiting bodies may be visible. Affected potatoes cannot be eaten

POTATO SILVER SCURF *HELMINTHOSPORIUM SOLANI*

PLANTS AFFECTED Potatoes, mainly in storage
SEASON Late summer to winter
WATCH FOR Pale brown and silvery marks on skin
ACTION NEEDED? No

Affected potato tubers develop pale, silvery spots, often surrounded by patches of light brown colour. The eating quality of the potatoes is unaffected and the tubers are not at increased risk of developing secondary rots. Silver scurf is caused by a soil-borne fungal infection, and although symptoms can sometimes be found at harvest, they only usually develop during storage, particularly where conditions are humid. No control measures are necessary.

The skin of affected potatoes may eventually flake off (scurf).

POTATO SPRAING *TOBACCO RATTLE VIRUS*

PLANTS AFFECTED Potatoes
SEASON Summer and autumn
WATCH FOR Corky growth on surface, brown bands in flesh
ACTION NEEDED? Yes, to prevent spread

This virus is spread by tiny eelworms in the soil. Affected potatoes may have raised, corky growths on their skin, and when the tubers are cut, curved bands of brownish discolouration are clearly visible in the flesh. The leaves of plants may also show some yellow mottling. Remove affected plants and weed regularly to prevent viruses spreading. Rotate crops and do not plant tubers that may be infected; always buy seed potatoes from reputable sources.

A band of discolouration inside a potato tuber is characteristic of potato spraing.

POWDERY POTATO SCAB *SPONGOSPORA SUBTERRANEA*

PLANTS AFFECTED Potatoes
SEASON Summer
WATCH FOR Scabby, brown dents in skin filled with powdery spores
ACTION NEEDED? Yes, to prevent spread

Small, brown patches form on the skins of affected potatoes, expanding to form dents filled with masses of dark brown, powdery spores. Infections can lead to severe damage. This fungal disease is more prevalent in heavy soils and wet growing seasons. Incorporate organic matter to improve drainage and practise crop rotation to help prevent infection. Remove infected tubers and do not compost. Avoid growing potatoes on the same site for at least three years.

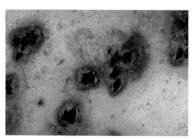

Powdery scab can be mistaken for common scab, but its scabs tend to be smaller, and rounder in shape.

VIOLET ROOT ROT *HELICOBASIDIUM PURPUREUM*

PLANTS AFFECTED Many plants
SEASON Spring to autumn
WATCH FOR Roots covered in purple fungal growth and may rot
ACTION NEEDED? Yes, to limit spread

The roots and storage organs (tubers and rhizomes) of plants become coated in thread-like, dark purple fungal growth. As the roots turn brown and start to rot, the fungus forms dark seed-like resting bodies, which drop into the soil. They can remain viable for years. Check the roots of plants with yellowing foliage and weak growth, and remove them immediately if symptoms are found. Improve drainage, as the disease is more common in wet soils. No chemical controls are available.

Soil tends to stick to affected carrots when they are unearthed.

BORON DEFICIENCY

PLANTS AFFECTED Root vegetables
SEASON Summer and autumn
WATCH FOR Split or rotten root crops
ACTION NEEDED? Yes, to correct deficiency

Discoloured and perhaps deformed foliage of root crops may be a sign of boron deficiency. A lack of boron can cause the roots of radishes and carrots to split, and the flesh of swedes, beetroots, and turnips to develop brown rings. These spoil the vegetable and can lead to rotting. Boron availability is limited on dry or alkaline soils, so keep plants well watered and avoid excessive soil liming. Apply borax to plots before sowing, mixed with horticultural sand to ensure an even spread.

Rings of brown discolouration inside roots are a symptom of boron deficiency.

CARROT CAVITY SPOT

PLANTS AFFECTED Carrots, parsnips
SEASON Summer and autumn
WATCH FOR Small, brown, oval marks on roots
ACTION NEEDED? Yes, to improve growing conditions

Calcium deficiency may lead to small, brown, oval marks on the surface of carrot roots – a disorder known as cavity spot. Foliage remains healthy and symptoms are only apparent when roots are lifted, when it is too late to take any action. Calcium deficiency is most likely a result of an inconsistent water supply during growth. Improve the soil with plenty of organic matter to help retain moisture, water during dry weather, and lime acidic soils to increase availability of calcium.

Brown patches may develop into lesions that are vulnerable to secondary fungal infections.

FROST DAMAGE

PLANTS AFFECTED Many plants
SEASON Autumn to spring
WATCH FOR Wilting, foliage and stem die-back
ACTION NEEDED? Yes, to protect vulnerable plants

Where very cold conditions persist, soil freezes and roots cannot reach liquid water, causing even hardy plants to wilt and die back. Extreme cold can also kill fine roots, resulting in a lack of vigour, or even death, when plants begin to grow in spring. Move containers under cover or insulate with bubble wrap. Apply a thick mulch of compost to soil in autumn to protect vulnerable plants. Lift tender plants like dahlias once they have died down and store somewhere frost-free for winter.

Plants grown in containers are more susceptible to cold weather damage.

PHYSICAL DAMAGE

PLANTS AFFECTED Any plant
SEASON All year
WATCH FOR Severed roots, damaged bulbs
ACTION NEEDED? Sometimes, to help prevent further problems

Physical damage to roots and bulbs can occur while working in borders with tools like forks and spades. Most plants will shrug off minor root damage, but punctured bulbs will be vulnerable to fungal infections and should be removed. Mark the position of bulbs when foliage dies down and explore soil around trees and shrubs gently before digging. High winds can also blow over trees and shrubs, tearing their roots as they fall. Avoid this by staking newly planted trees securely.

Be careful when harvesting potatoes as it's easy to prong them on a garden fork.

POTATO HOLLOW HEART

PLANTS AFFECTED Potatoes
SEASON Summer and autumn
WATCH FOR Cavities at the centre of tubers
ACTION NEEDED? Yes, to prevent damage in subsequent seasons

Seemingly normal potato tubers sometimes have rounded or star-shaped cavities at their centre when cut open. If this occurs, a large proportion of the potato crop is likely to be affected. The exact cause of this disorder is unclear, but it is thought to result from inconsistent growing conditions, such as heavy watering after dry weather. Prevent it recurring by incorporating organic matter into the soil to help retain moisture and watering crops regularly as tubers develop.

The tissue around the hollow heart is brown, but the rest of the tuber is perfectly edible.

ROOT FANGING

PLANTS AFFECTED Carrots, parsnips
SEASON Summer and autumn
WATCH FOR Forked roots
ACTION NEEDED? Sometimes, to prevent symptoms in future

The roots of carrots and parsnips sometimes fork and develop multiple growing points. This doesn't affect their eating quality, but makes roots more awkward to clean and peel. Fanging is caused when a root's tip is physically damaged by stones in the soil, or when the nutrient content of the soil is too high (for example, when soil been recently manured). Remove stones where possible or grow root crops in raised beds. Add manure to beds in autumn and avoid overuse of fertilizers.

Forked carrots may look odd but still taste delicious.

ROOT SPLIT

PLANTS AFFECTED Root crops
SEASON Summer and autumn
WATCH FOR Long cracks in harvested roots
ACTION NEEDED? Sometimes, to prevent symptoms in future

Root crops may develop deep cracks or splits in their skin, which provide an access point for secondary infections that can cause rotting. Inconsistent water supply during development is a common cause and can be avoided by adding organic matter to the soil every year to retain moisture, and watering crops regularly. Choose varieties with resistance to splitting. Late crops may also split when exposed to frost, so lift and store carrots and beetroot before cold weather in autumn.

Drought followed by heavy rain is a common cause of root split in root crops.

WATERLOGGING

PLANTS AFFECTED Any plant
SEASON All year
WATCH FOR Wilting, yellowing foliage, leaf fall
ACTION NEEDED? Yes, to improve drainage

Waterlogged soil will block the movement of oxygen to roots, which turn black and die, leaving plants unable to take up water and producing symptoms similar to drought. Above ground, the plant's leaves wilt, turn yellow, and fall. Improve drainage by adding plenty of organic matter and grit to open out the structure of heavy clay soils: rapid action may save plants that have been waterlogged. Ensure containers have drainage holes before planting and avoid overwatering plants.

Overwatered pot plants like this cyclamen rapidly yellow, die, and rot away.

Stems, such as this *Cornus sanguinea* 'Anny's Winter Orange' provide structural interest in the winter months.

STEMS

Stems are not just there to support plants. They transport water and nutrients to foliage and flowers, which makes them a target for sap-sucking insects. Stem damage can have serious consequences for a plant's health. Wilting or dieback can be signs of problems at the roots, but many diseases will also infect stems directly, so be alert for discolouration and signs of fungal growth as well as deformities and roughened patches of bark.

APHIDS APHIDOIDEA

PLANTS AFFECTED Many plants outdoors and in greenhouses
SEASON Spring to autumn
WATCH FOR Insects, deformed shoot tips, honeydew/sooty mould
ACTION NEEDED? Yes, to prevent serious infestations

Aphids are a very common pest. You are likely to see clusters of these green, black, or pinky-brown insects sucking sap from stems as well as leaves (see p.79), especially on new, soft growth at the shoot tips. Check any shoots showing signs of distortion and watch out for sooty mould on upper leaf surfaces, which relies on the honeydew secreted by aphids. Encourage beneficial wildlife into the garden to control aphid numbers and squash any clusters that you find.

Aphids only become a serious problem if present in large numbers.

ASPARAGUS BEETLE CRIOCERIS ASPARAGI

PLANTS AFFECTED Asparagus
SEASON Spring to autumn
WATCH FOR Bark damage, stems dying back
ACTION NEEDED? Yes, to control numbers

Affected asparagus stems turn dry and yellow-brown, and the feathery foliage may also be eaten (see p.80). The damage is caused by adult asparagus beetles (whose dark bodies are punctuated by large cream spots) and by their larvae. Be vigilant from spring and pick off any beetles or grubs that you find to prevent their numbers increasing. Organic insecticides containing pyrethrum can also help. Cut down and burn affected stems in autumn to prevent beetles overwintering inside.

The larvae are around 8mm (⅓in) long with slug-like bodies and black heads.

BLACKCURRANT BIG BUD MITE

CECIDOPHYOPSIS RIBIS

PLANTS AFFECTED Blackcurrants
SEASON Winter to spring
WATCH FOR Unusually swollen buds, leaves fail to emerge
ACTION NEEDED? Sometimes, to control spread

In late winter, buds on blackcurrant stems may swell, becoming spherical rather than long and slender. These buds dry out or produce stunted foliage and few flowers, leaving stems bare. Each bud contains hundreds of tiny gall mites, which feed on sap and cause the deformed growth. Pick off affected buds promptly and remove damaged plants. Buy certified disease-free plants or grow the resistant variety 'Ben Hope'.

Mites emerge from affected buds in early summer and seek out new plants to colonize.

CUCKOO SPIT PHILAENUS SPECIES

PLANTS AFFECTED Many plants
SEASON Late spring and summer
WATCH FOR Blobs of white foam on stems
ACTION NEEDED? No

Small masses of white froth on the stems of plants, often at junctions with leaf stalks, have no connection with cuckoos. Within the froth is a single yellow-green insect, the nymph of a froghopper, which feeds on sap. Plants can usually tolerate a low level of damage and adult froghoppers tend to disperse around the garden, so cuckoo spit can generally be left alone. However, you should remove it from young shoots, where damage caused by the feeding nymph may distort new growth.

Cuckoo spit is so named as it appears in spring when the cuckoo's call is also heard.

DEER CERVIDAE

PLANTS AFFECTED Many plants
SEASON All year
WATCH FOR Young shoots and tree bark eaten
ACTION NEEDED? Yes, to protect vulnerable plants

Deer cause considerable damage where they venture into gardens, biting and pulling away young shoots during spring and summer (usually leaving a distinctive ragged edge on one side of stems), and eating the bark of woody plants in winter. To fully protect plants deer need to be excluded with a strong, 2m- (6ft-) high wire mesh fence and secure gates. Where this is not practical, enclose individual beds with chicken-wire cages while they establish and protect saplings with tree guards.

Sonic deterrents can work against deer but must be moved regularly to stop animals from becoming accustomed to the sound.

RABBITS LEPORIDAE

PLANTS AFFECTED A wide range of plants
SEASON All year
WATCH FOR Soft growth grazed and bark gnawed
ACTION NEEDED? Yes, to protect plants

Rabbits have a big appetite for leaves and vegetables (see p.85), but will also eat new shoots of trees and shrubs if they are within reach, and gnaw bark from woody plants, killing them if the circumference of the trunk is stripped. Exclude rabbits with a wire mesh fence at least 1.2m (4ft) high and sunk 30cm (12in) below soil level. Ensure that gates are also rabbit-proof and kept closed. Protect plants or beds with netting or wire mesh cages and young trees with spiral tree guards.

Wrap rabbit guards around the bottom 60cm (24in) of trunk, to protect bark from damage.

PHLOX EELWORM DITYLENCHUS DIPSACI

PLANTS AFFECTED Phloxes, both annual and perennial
SEASON Spring to summer
WATCH FOR Short, swollen stems with narrow leaves
ACTION NEEDED? Yes, to prevent spread

These tiny nematode worms overwinter in dormant phlox buds and emerge in spring. Affected plants are stunted, their stems swell and they may split near the base; leaves may become narrower and deformed near the shoot tip. No treatments are available. Remove infected plants and burn or dispose of them in domestic waste to prevent spread. Phlox eelworms are not found on roots; propagate healthy stock from root cuttings of infected plants and plant away from the infected area.

Eelworms feed inside stems, damaging tissue and causing distorted, discoloured growth.

SMALL ERMINE MOTH YPONOMEUTA SPECIES

PLANTS AFFECTED *Prunus, Crataegus, Euonymus, Sedum*
SEASON Spring to summer
WATCH FOR Stems draped with webbing, defoliation
ACTION NEEDED? Sometimes, to protect smaller plants

The caterpillars of the small ermine moth feed on leaves beneath a canopy of fine, silky white webbing that they themselves produce. The larvae hatch in late summer and overwinter on plants before beginning to feed in spring. They can usually be tolerated, but pruning out affected stems promptly from smaller plants will help limit defoliation. If necessary, spray with contact insecticides, such as pyrethrum or synthetic pyrethroids.

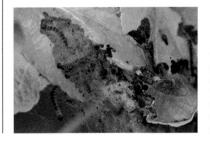

The caterpillars are creamy-white with black spots. They are clearly visible under their webbing on this apple tree.

SCALE INSECTS VARIOUS SPECIES

PLANTS AFFECTED Trees, shrubs, woody climbers
SEASON All year
WATCH FOR Small bumps on stems, sooty mould on leaves
ACTION NEEDED? Yes, to prevent infestation

Clusters of small, waxy shells are sometimes visible on the stems of plants as well as leaves (see p.87). These sap-sucking scale insects may weaken plants and also secrete sticky honeydew, which attract ants and fuels development of black sooty mould. Beneficial wildlife may help to control pest numbers and scales can be picked off by hand. Adults are protected from chemical controls by their shells, but spraying with contact insecticides in midsummer kills immature nymphs.

Scale insects favour plants grown in sheltered positions and against walls.

SLUGS AND SNAILS GASTROPODA

PLANTS AFFECTED Young plants, and new herbaceous shoots
SEASON Spring and summer
WATCH FOR Wilting, shoots dying back, seedlings collapsing
ACTION NEEDED? Yes, to protect vulnerable plants

The soft stems of new spring and summer growth are particularly vulnerable to damage by slugs and snails. Stems may be totally or partially severed, causing seedlings or shoots to collapse or die back. Damaged stems can also affect growth and allow in secondary infections. Protect seedlings and new plants with cloches or other barriers, or by applying a garlic drench regularly. You can also use traps and manual collection to reduce the numbers (see pp.88–89).

Damage to stems is often carried out by smaller slugs that emerge from the soil.

SQUIRRELS SCIURUS SPECIES

PLANTS AFFECTED Trees
SEASON Mainly spring to summer
WATCH FOR Bark stripped from trunk or branches
ACTION NEEDED? No, as little can be done

Agile squirrels may strip bark from trunks and branches of established trees, but usually leave saplings untouched. These injuries are an entry point for infections and the growth above the damage will die back if bark has been removed around a whole trunk or branch. Nothing can be done to protect large trees or keep grey squirrels out of gardens. As a non-native species in the UK, grey squirrels cannot be trapped and then released, so must either be humanely killed or tolerated.

Trees can tolerate some bark removal, but will die if the damage is extensive.

WOOLLY APHIDS ERIOSOMA LANIGERUM

PLANTS AFFECTED Apples, crab apples, Cotoneaster, Pyracantha
SEASON Spring to autumn
WATCH FOR Fluffy, white patches on bark
ACTION NEEDED? Yes, to limit numbers

Patches of a fuzzy, waxy white secretion may appear around split bark and old pruning cuts in spring, and then on young branches in summer. These coverings are made by sap-sucking aphids beneath. Their activity causes swellings on young branches, which may split open. Birds, ladybirds, and earwigs will help control aphid numbers, and colonies can be scrubbed off with a stiff brush or treated with insecticides or plant invigorators. Winter washes are not effective as nymphs hide deep in bark.

Damage caused by woolly aphids may provide access points for secondary infections.

ASH DIEBACK *HYMENOSCYPHUS FRAXINEUS*

PLANTS AFFECTED Ash trees (*Fraxinus* species)
SEASON Spring to autumn
WATCH FOR Expanding dark spots on bark, branches dying back
ACTION NEEDED? Yes, to report infection

This fungal disease is lethal to infected trees; young trees may be killed in a single season. Dark oval patches form on the bark, expanding quickly to form lesions on trunks and branches. Where this damage rings the stem the growth above it will die back. Leaves may also develop dark brown patches. Spores spread on the wind and infect trees through their leaves. Infection cannot be prevented or treated, but should be reported to the relevant plant health authority in the area.

The bark on this affected sapling has cracked revealing discoloured tissue beneath.

BACTERIAL CANKER *PSEUDOMONAS* SPECIES

PLANTS AFFECTED Cherry trees and other *Prunus* species
SEASON Spring and summer
WATCH FOR Sunken patches of bark with resinous ooze
ACTION NEEDED? Yes, to prevent spread

In spring, trees infected with bacteria may develop abnormal areas of bark that collapse to form dents on the trunk or on branches. Sticky, amber-coloured ooze appears at these sites, and new shoots may fail to open or die back suddenly. Small brown spots appear on leaves, forming holes as the dead tissue falls away (see *shothole p.97*). Prune infected growth back to healthy wood in summer to prevent spread and burn or bin prunings. Grow varieties of cherries or plums with resistance.

Reddish-brown ooze on this apricot tree trunk indicates infection by *Pseudomonas*.

BRACKET FUNGUS VARIOUS SPECIES

PLANTS AFFECTED Many trees
SEASON All year
WATCH FOR Large bracket-shaped fungi on trunk or branches
ACTION NEEDED? Yes, to assess stability of infected trees

These large, flat-topped growths on trunks or main branches of trees are the spore-producing "fruiting" bodies of fungi. Strands of the fungus within the tree decay its heartwood and weaken its structure, eventually causing branches to drop or the whole tree to fall; removing the brackets will not prevent this. Little can be done to stop or treat infection, other than avoiding unnecessary pruning. Bring in an arboriculturalist to assess affected trees and make them safe.

The heartwood of a tree is likely to have sustained significant damage by the time the fungal fruiting bodies appear.

CLEMATIS WILT *CALOPHOMA CLEMATIDINA*

PLANTS AFFECTED Clematis, especially large-flowered varieties
SEASON Spring to autumn
WATCH FOR Wilting stems, leaves, and buds
ACTION NEEDED? Yes, to control spread

Clematis shoots infected with this fungus wilt suddenly from the tips. New growth becomes floppy and any woody stems develop dark discolouration inside. This fungal infection usually causes symptoms in just a few stems, so can be distinguished from wilting caused by drought or root diseases, which affects the whole plant. Cut all diseased stems back to healthy, unmarked growth (which might be below soil level) and sterilize your pruning tools to prevent cross-contamination.

Leaves and flowers on affected stems will wilt and die back.

CORAL SPOT *NECTRIA CINNABARINA*

PLANTS AFFECTED Woody plants
SEASON All year
WATCH FOR Small orange-pink pustules on dead wood
ACTION NEEDED? Yes, to prevent infection and spread

This fungus infects plants through wounds caused by physical injury or poor pruning and spreads through the stems. Tiny coral-pink pustules appear on the bark of dead wood. Coral spot is a weak pathogen and will cause serious damage only on plants that are already struggling. Prevent infection by making tidy pruning cuts with clean, sharp tools. Prune out any dead wood showing signs of coral spot immediately and regularly clear away any stems that have fallen to the ground.

The pustules produce spores that are spread mainly in wind-blown rain.

CROWN GALL *RHIZOBIUM RADIOBACTER*

PLANTS AFFECTED Many woody and herbaceous plants
SEASON All year
WATCH FOR Knobbly swellings on stems and roots
ACTION NEEDED? Yes, to prevent spread

This bacterial disease causes clusters of rounded swellings, known as galls, to develop at the base of stems or occasionally further up the stem or trunk. Large galls may obstruct the flow of water and nutrients to foliage, causing parts of the plant to die back. Remove infected plants to avoid contamination and grass over (or grow potatoes in) the areas where the plants grew for at least one year to reduce the likelihood of subsequent infections.

The bacteria that cause crown gall enter plants through wounds.

DAMPING OFF *VARIOUS FUNGI*

PLANTS AFFECTED Seedlings
SEASON Mainly spring
WATCH FOR Seedlings fail to emerge or collapse
ACTION NEEDED? Yes, to prevent problems

Damping off is a condition that causes seedlings to die. They collapse as their stems rot at the base and may then be engulfed in fuzzy, white fungal growth. The fungal pathogens responsible can spread quickly in infected soil or water, resulting in the loss of large patches of seedlings in a pot or tray, or the failure of seedlings to emerge if they are affected as they germinate. Damping off commonly occurs during spring, because the warm, humid conditions provided to raise seedlings under cover are also ideal for fungal growth.

CONTROL Prevent the introduction of these fungal pathogens by using sterilized compost and mains water when raising seedlings. If rainwater must be used, ensure that your water butt is covered to prevent contamination with organic material that could carry fungal spores. Clean all pots, seed trays, propagators, and the greenhouse with a disinfectant before use. Sow seeds thinly and remove any covers on propagators or pots as soon as germination takes place.

Densely sown seedlings grown in containers lacking adequate drainage holes are susceptible to damping off.

DIEBACK VARIOUS

PLANTS AFFECTED Many plants
SEASON All year
WATCH FOR Stems turning brown, foliage dying off
ACTION NEEDED? Yes, to find cause and prevent spread

Dieback is a condition, rather than a particular infection. Affected stems turn brown or darken as their tissue dies and leaves wilt and die. The damage may begin at stem tips, at the bases of stems, or anywhere along their length, and can kill the entire plant if symptoms spread. Dieback can be initiated by many fungal infections, including grey mould (*Botrytis*) and spur blight, and (less often) bacterial diseases like fireblight. Root rots, which prevent plants taking up water, can cause dieback, as can waterlogging and drought, so check soil conditions and root health where no cause is apparent above ground. Frost or wind damage may kill shoot tips in cold winters. Severe defoliation by insects and poor pruning can also result in dieback.

CONTROL Try to identify the cause to prevent it spreading or recurring. Prune out all dead stems, cutting them back to just above a healthy bud with clean, sharp tools. Provide better growing conditions by watering and mulching dry soil or improving drainage. Avoid overcrowding plants, prune

Dieback of stems on blackcurrant bushes is often caused by fungal infections.

branches of trees and shrubs to maintain an open shape, and ventilate greenhouses to create good air circulation and prevent fungal infections. Protect vulnerable plants from frost by covering with fleece.

FOOT AND ROOT ROT VARIOUS FUNGI

PLANTS AFFECTED Mainly soft-stemmed annual and herbaceous plants
SEASON Spring to autumn
WATCH FOR Wilting shoot tips, darkening at stem base
ACTION NEEDED? Yes, to prevent spread

Various fungal infections can rot tissues at the foot of the stem, causing them to darken and sometimes soften, and may spread to roots. Infection is common in greenhouse plants and usually spreads via contaminated pots, compost, or water. Maintain good garden hygiene and use sterilized compost for planting in containers, and water with tap water. Remove infected plants immediately to prevent infection spreading.

The pathogens that cause foot rot can survive for many years in the soil.

FORSYTHIA GALL *PHOMOPSIS* SPECIES

PLANTS AFFECTED Forsythia shrubs
SEASON All year
WATCH FOR Roughened swellings on stems
ACTION NEEDED? No

Swollen galls may develop on the stems of forsythia shrubs. Each one is roughly spherical and up to 1.5cm (⅝in) in diameter, with a knobbly surface that gives the impression of many small growths clustered together. Their cause is uncertain, but because they rarely affect the growth or vigour of plants they are usually easily tolerated. Where they do spoil a plant's appearance, stems disfigured by galls should be pruned back into unaffected growth or cut out entirely.

Forsythia galls may appear on isolated stems or more widely across a plant.

FUNGAL CANKER *NECTRIA GALLIGENA*

PLANTS AFFECTED Apple and pear trees
SEASON All year
WATCH FOR Sunken areas of discoloured, cracked bark; swelling
ACTION NEEDED? Yes, to prevent problems

Infection by this fungus causes circular areas of bark to sink inwards. Small branches may be ringed and killed rapidly, while larger branches develop perennial cankers with rings of flaking bark, raised swollen edges and sometimes exposed wood at their centre. These weaken and eventually kill the branch. Prune out smaller branches and cut away all damaged tissue from larger cankers before treating with a wound paint to prevent reinfection. Grow resistant apple varieties if possible.

Remove branches affected by fungal canker back to a healthy limb.

GREY MOULD *BOTRYTIS CINEREA*

PLANTS AFFECTED Many, especially in greenhouse
SEASON All year
WATCH FOR Grey fuzzy mould, softening stems
ACTION NEEDED? Yes, to prevent spread

Botrytis infects plants through wounds or flowers, particularly in wet weather, when air circulation is poor, or if plants are overcrowded. Stems on non-woody plants develop brown patches, soften, and become covered in pale grey fungal growth. Foliage and flowers above the affected area wilt, yellow and die, and the whole stem will often collapse. Other parts of the plant may also develop fuzzy mould. Cut out infected stems immediately and tidy up dead or damaged plant material regularly.

Cleaning and disinfecting greenhouses regularly helps keep minimize infection.

HONEY FUNGUS *ARMILLARIA* SPECIES

PLANTS AFFECTED Trees, shrubs and woody perennials
SEASON All year
WATCH FOR Toadstools in late summer to autumn, distinctive mushroomy smell, plants dying suddenly, smaller leaves
ACTION NEEDED? Yes, to reduce spread

Honey fungus affects a very wide range of plants, with some, such as willow, rhododendron, buddleja, and lilac, particularly vulnerable. Plants may die suddenly, particularly during hot, dry weather, may fail to come into leaf in spring, or die back gradually over several years as a result of root damage. To confirm infection by honey fungus, look for cracks in the bark just above soil level and a layer of white fungal growth with a strong mushroomy scent beneath the bark at the base of stems. Clumps of honey-coloured toadstools may also emerge around affected plants in autumn, but can be left as their spores are not responsible for infections. Honey fungus spreads via root contact between plants or tough, black, bootlace-like strands, called rhizomorphs, which grow through the soil to infect the roots of new hosts up to 30m (100ft) away.

CONTROL There are no chemical measures effective against honey fungus. Dig out and destroy infected plants, including

The fruiting bodies of honey fungus may be up to 15cm (6in) high and 2cm (¾in) wide.

their stumps and roots, as soon as possible. This will remove the fungus's food source and help to prevent rhizomorphs spreading through the soil. Replant affected areas with plants that show resistance. Provide good growing conditions, as healthy plants are better able to resist infection.

LEAFY GALL *RHODOCOCCUS FASCIANS*

PLANTS AFFECTED Annuals and herbaceous perennials
SEASON Spring and summer
WATCH FOR Dense, distorted leaf growth near stem base
ACTION NEEDED? Yes, to prevent spread

When *Rhodococcus* bacteria infect a wound on a stem, they interfere with the hormones that regulate plant development causing the growth of deformed leaf masses, usually close to the base of a stem. Affected shoots tend to be stunted with small, thickened, or fasciated leaves. Where symptoms can't be tolerated, remove affected plants and their surrounding soil and dispose of them in your domestic waste. Wash your hands and any tools used to prevent the spread of infection.

Clusters of deformed leaves on this pelargonium stem are a clear sign of bacterial infection.

PEONY WILT *BOTRYTIS PAEONIAE*

PLANTS AFFECTED Herbaceous and tree peonies
SEASON Spring and summer
WATCH FOR Stems, leaves, and flower buds wilt and turn brown
ACTION NEEDED? Yes, to prevent spread

This fungus affects peonies, producing brown patches on leaf and flower stems, which collapse and die, giving plants a wilted appearance. Flower buds also turn brown and fail to open (see p.107). Cut back infected stems into healthy growth. Do not compost the waste. Instead, burn it or send it to landfill to prevent infectious spores being released, otherwise black resting bodies (sclerotia) will form, overwinter in the soil, and reinfect plants the following spring. No chemical treatments are available.

A fuzzy grey mould may be seen on infected areas during damp weather conditions.

POTATO BLACKLEG *ERWINIA CAROTOVORA*

PLANTS AFFECTED Potatoes
SEASON Summer
WATCH FOR Yellow, stunted foliage, black rot on stem bases
ACTION NEEDED? Yes, to prevent spread

Affected potato plants grow slowly and their leaves yellow, curling in on themselves. Distinctive black stem rot begins at soil level and spreads upwards, causing plants to collapse. Remove infected plants immediately to prevent disease spreading. Grow resistant varieties and rotate crops to avoid pathogens building up in the soil. Improve drainage, as the disease thrives in wet soil. If storing your own seed potatoes, lift and keep them in dry conditions to stop any bacteria spreading.

The bacteria responsible for infection are usually introduced by planting symptomless tubers.

RASPBERRY CANE BLIGHT *PARACONIOTHYRIUM*

PLANTS AFFECTED Raspberries
SEASON Summer
WATCH FOR Canes that die back, withering foliage
ACTION NEEDED? Yes, to prevent spread

Raspberry canes infected with this fungus start to die back when they are in full leaf in summer. The bases of canes turn dark brown, split open, and become brittle so that they are easily snapped. Tiny, black, bead-like fruiting bodies form on dead wood and produce fungal spores. Cut out infected canes below soil level to prevent spread and disinfect secateurs between cuts. Prune during dry weather as fungal spores are readily spread by water splash.

Foliage withers rapidly where raspberry canes have been infected through injuries.

SMUTS VARIOUS FUNGI

PLANTS AFFECTED Alliums, anemones, winter aconites, and others
SEASON All year
WATCH FOR Brown swellings rupturing to release powdery spores
ACTION NEEDED? Yes, to prevent spread

A wide range of plants is affected by smuts – fungi that cause dark, blister-like swellings to appear along stems and sometimes leaves. The swellings burst to release thousands of sooty spores. In some plants it is possible to cut out the affected areas, but otherwise remove infected plants promptly and dispose of them in domestic waste. Avoid spore dispersal because smut spores can persist in soil for many years. Don't plant closely related plants on an infected site.

Affected stems are often distorted and their growth is significantly reduced.

SPUR BLIGHT MONILINIA LAXA

PLANTS AFFECTED Apple, pear, apricot, peach, and *Prunus* trees
SEASON Spring and summer
WATCH FOR Twigs dying back and developing pale brown pustules
ACTION NEEDED? Yes, to prevent spread

On some trees, fruit grows on short, branching shoots known as spurs. Spur blight causes these fruiting spurs to die back, and foliage to wilt and wither. Dead wood may develop small, rounded, pale brown fungal pustules. Spur blight is associated with blossom wilt (see p.107), which causes blossom to wither and die in spring. Prune out infected spurs, cutting back into healthy wood, and remove any fruits showing symptoms of brown rot (see p.114) in autumn to prevent infection spreading.

Pustules on affected spurs release fungal spores that will spread the infection.

VERTICILLIUM WILT VERTICILLIUM SPECIES

PLANTS AFFECTED A wide range of edible and ornamental plants
SEASON All year, but symptoms usually occur in summer
WATCH FOR Wilting and yellowing foliage, branch dieback
ACTION NEEDED? Yes, to prevent spread

The foliage of plants affected by this soil-borne fungal disease suddenly wilts from the base while the rest of the plant appears healthy. Infection spreads from the roots through the water-carrying tissues in stems, where it is visible as dark brown staining. There is no treatment, so remove infected plants, their roots, and surrounding soil immediately. Weed regularly, as some weeds also harbour the pathogen. Avoid replanting with vulnerable plants such as *Acer* and *Cotinus*.

This infected cherry laurel has yellow, withering leaves and stem dieback typical of verticillium wilts.

WITCHES' BROOM VARIOUS

PLANTS AFFECTED Trees and shrubs
SEASON All year
WATCH FOR Dense twiggy growths among branches
ACTION NEEDED? Yes, where they spoil a plant's appearance

Dense clusters of slender twigs, known as witches' brooms, appear among the branches of deciduous trees and shrubs. They are clearly visible when branches are bare in winter, but far less so when plants are in leaf. This abnormal growth results from fungal infections or damage caused by insect pests. Nothing can be done to prevent witches' brooms forming, but they can be removed where they spoil the appearance of a plant by pruning affected branches back to healthy growth.

Witches' brooms can make an interesting curiosity in large trees like this birch.

BORON DEFICIENCY

PLANTS AFFECTED Mainly vegetable crops
SEASON Spring to autumn
WATCH FOR Roughened or hollow stems, stunted growth
ACTION NEEDED? Yes, to correct deficiency

Boron is essential for healthy growth, especially in brassicas. Insufficient boron causes roughened, hollow stems and leaf stalks in cabbages, cauliflowers, and broccoli, and brown cracks in celery stalks. Growth may be stunted, with yellowing lower leaves; root crops can also be affected (see p.128). Boron availability is limited on dry, alkaline soils, so keep plants well watered and avoid excessive liming. Apply borax mixed with horticultural sand to ensure it is spread evenly before sowing.

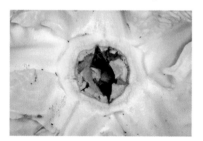

A lack of boron leads to hollowed-out stems in cabbages. Here, stem damage has allowed rot to take hold.

FROST DAMAGE

PLANTS AFFECTED Many plants
SEASON Autumn to spring
WATCH FOR Brown or black foliage at shoot tips
ACTION NEEDED? Sometimes, to provide protection

Frost may cause soft stem tips and foliage to turn brown or black, take on a watery appearance, and eventually die back. Tender plants and those growing in frost pockets are most susceptible. Often, only the exposed parts of a plant are affected. Prune out any damaged growth in late spring. Avoid clipping shrubs and using high-nitrogen fertilizers in autumn, because the new growth is easily frosted. Protect vulnerable plants with horticultural fleece covers during cold weather.

These tender new laurel stems and leaves have been damaged by frost.

DROUGHT

PLANTS AFFECTED Most plants, especially those in containers
SEASON Mainly spring and summer
WATCH FOR Wilting of all stems and foliage and poor growth
ACTION NEEDED? Yes, to prevent soil drying out

The stems and foliage of herbaceous plants will wilt from the tips when the soil is dry, especially when the weather is hot. Prolonged drought will result in poor growth and stems dying back from the tips, even on woody plants. Regular watering of container plants and those grown under cover or on sandy or chalky soils will keep the soil moist and growth vigorous. Incorporate organic matter into light soils to retain moisture and apply a thick spring mulch to prevent evaporation.

Drought-stressed plants can quickly be revived with a thorough watering.

IRREGULAR WATERING

PLANTS AFFECTED Trees and shrubs
SEASON All year
WATCH FOR Splits or cracks in bark
ACTION NEEDED? Yes, to help retain soil moisture

Erratic growth that results from an irregular supply of water at the roots can cause cracks or splits to appear along woody trunks, stems, or branches. These cracks often heal naturally, becoming raised at the edges; however, while the plant heals, it is susceptible to secondary infections. Bark cracks are most common in plants grown in containers or on light soils, but also occur when drought is followed by heavy rain. Add organic matter to aid water retention in soil, and water consistently.

Cracks may sometimes be sufficiently deep to expose the wood beneath.

POOR PRUNING

PLANTS AFFECTED Trees and shrubs
SEASON All year
WATCH FOR Dieback, wounds failing to heal, secondary infections
ACTION NEEDED? Yes, to remove damaged wood where possible

Woody plants are easily damaged by poor pruning. Badly placed cuts will not heal properly or may leave a stub that will die back. Untidy cuts made with blunt tools will also heal poorly. These injuries invite secondary infections, which cause wood to die back and rot, and can spread through the plant. Prevent problems by using clean, sharp tools to cut just above a healthy bud or at the swelling, known as the shoulder, where a branch meets the trunk. Prune out any dead wood.

This stub, left when a branch was cut too far from the trunk, is starting to die back.

WATERLOGGING

PLANTS AFFECTED Many plants
SEASON All year
WATCH FOR Stems and leaves wilting and dying back
ACTION NEEDED? Yes, to improve soil drainage

Waterlogged plants show similar symptoms to those affected by drought: their stems and leaves wilt and then die back from the tips downwards. Waterlogging occurs where soil is heavy and poorly drained or containers lack drainage holes. Deprived of oxygen, the roots die, leaving plants unable to take up water. Wet conditions also encourage rots at the base of stems. Ensure that containers have drainage holes and add organic matter and grit to soils to improve drainage.

Plants will wilt very rapidly when the soil at their roots is waterlogged

WEEDKILLER DAMAGE

PLANTS AFFECTED All plants
SEASON Spring to autumn
WATCH FOR Deformed or stunted stems and leaves
ACTION NEEDED? Yes, to prevent damage

Plants are damaged if fine weedkiller spray drifts onto them on the wind, or if contaminated equipment is used to water them. Whole plants or isolated areas produce twisted or stunted stems, with deformed leaves and sometimes unusual swollen gall-like growths or proliferations of twiggy shoots. The plants may die back if symptoms are severe. Prune out damaged growth where possible and remove dead plants. Only use weedkillers where no other options are available.

These gall-like growths on a brassica stem were caused by accidental exposure to weedkiller.

WIND DAMAGE

PLANTS AFFECTED Trees, shrubs, climbers, and tall perennials
SEASON All year
WATCH FOR Stems bent or snapped after bad weather
ACTION NEEDED? Yes, to remove damaged growth

The branches of trees and shrubs may be damaged by high winds, while whippy climbers are easily broken if not secured to a support. Tall perennials can also be also blown over, especially if carrying heavy flowers or wet foliage. Prune damaged branches of woody plants back to a healthy bud or the trunk to promote rapid healing. Cut back broken stems and prop up those that remain with supports. Regularly check the integrity of plant supports and ties.

These tall artichoke plants have been blown over by strong winds.

INDEX

Bold text indicates a main entry for the subject.

Author Jo Whittingham

PUBLISHER ACKNOWLEDGMENTS

DK would like to thank Mary-Clare Jerram for developing the original concept, Margaret McCormack for indexing, Diana Vowles for proofreading, and Paul Reid, Marek Walisiewicz, and the Cobalt team for their hard work in putting this book together.

PICTURE CREDITS

The publisher would like to thank the following for their kind permission to reproduce their photographs:

Alamy Stock Photo: A Garden 20tr; Alec Scaresbrook 109br; Amelia Martin 6c; Andrew Harker 35bc; Andrew Hasson 134tl; Arterra Picture Library 29cl; Ashley Cooper pics 108br; Avalon.red 109tr, 125bl, 126tr, 134br; blickwinkel 21br, 31tr, 117tr, C. Avilez 107br; Cavan Images 33br; Christian Hütter 115tr; christopher miles 17tl, 20br; Clare Gainey 4-5l; Claudio Fichera 113bl; Colin Underhill 117br; Dave Bevan 86br, 92bl, 96tl, 112tl, 122tr, 129bl, 133tl; David Norton 43bl; Deborah Vernon 9br; Denis Crawford 42bl; Diane Randell 43t; Farmer 122tl, 126bl; Fcerez 49bl; FLPA 75br, 98bl, 100tl; Frank Hecker 105br, 120tr; gardeningpix 132br, 141tr; garfotos 128bl; Genevieve Vallee 18br; GFK-Flora 80br, 86tr, 109bl, 135tr, 136br; Gillian Pullinger 38bl; GKSflorapics 68cl; Graham Prentice 104t; Hhelene 133tl; James Davidson 48br; Jeanette Teare Garden Images 124tr; Jinny Goodman 129tl; Joe 94tl; Juniors Bildarchiv GmbH 106tl; katewarn images 31cl; Kathy deWitt 140tr; Larry Doherty 139br; Les. Ladbury 11br; Linda Jones 101br; Marcus Harrison 76c; Martien van Gaalen 34bl; Martin Hughes-Jones 121bl, 122bl; Matthew Taylor 41bl; Mike Ford 141br; nagelestock.com 74tr; Nature and Science 120br; Nigel Cattlin 9bl; 12cl, 42bc, 80tr, 82tr, 85tr, 90tr, 90bl, 92tl, 94bl, 97br, 99cr, 102tl, 102tr, 105tr, 109tl, 111br, 113tr, 116br, 119tl, 119tr, 121tl, 121br, 124bl, 125tl, 125br, 127tl, 127bl, 128tl, 128tr, 131tr, 131bl, 133br, 134tr, 136bl, 138tr, 138bl, 139tr, 140tl; Papilio 83tl; Paulo Oliveira 121tr; pbpvision 133bl; Philip Mugridge 105bl; RM Floral 46br; ronstik 47tr; Ros Crosland 32cl; Steven Chadwick 85br; Vaclav Mach 52br; Washington Imaging 126br; wda delta 140bl; Westend61 GmbH 50c; Wlodzimierz Dondzik 44br; Zoonar GmbH 132tl.

Dorling Kindersley: 123RF.com: Denis Tabler 31bc; 123RF.com: Marie-Ann Daloia / mariedaloia 122br; 123RF.com: schan 131tl; Igor Zhorov 8br; Alan Buckingham 13tr; 14cr, 15bc, 19bl, 48cl, 56tr, 57bc, 60br, 79bl, 85tl, 86tl, 87tr, 92br, 95br, 97tr, 97bl, 103bl, 111bl, 113tl, 113br, 114tr, 115bl, 117bl, 136tr; 137tl; Brian North / RHS Hampton Court Flower Show 2012 28tr; Brian North / Waterperry Gardens 93tr; David Fenwick 94tr; Dreamstime.com: Frankjoe1815 10bl; Dreamstime.com: Inna Kyselova 33bc; Dreamstime.com: Jordan Roper 10cr; Dreamstime.com: Marilyn Barbone 29tr; Dreamstime.com: Photographyfirm 29br; Dreamstime.com: Vitaliy Parts 123tr; Dreamstime.com: Whiskybottle 137br; Fotolia: Thomas Dobner / Dual Aspect 30br; Mark Winwood / Ball Colegrave 36cl, 78t; Mark Winwood / RHS Chelsea Flower Show 2014 108bl; Mark Winwood / RHS Wisley 64tr, 66tr, 68tr, 106tr, 130t; Neil Fletcher 21bl; Peter Anderson 8cl, 17bc, 24tr, 66bl; Thomas Marent 28br.

GAP Photos: Dave Bevan 124br; Lee Avison 22c; Thomas Alamy 139bl.

Getty Images: 3drenderings 120tl; Akchamczuk 138br; coramueller 11tl; Goodboy Picture Company 40c; jess311 90tl; Magdevski 2c; SolStock 9tl.

Jo Whittingham: 140br.

Cover images: Front: Alamy Stock Photo: Paula French
Illustrations by Cobalt id.

All other images © Dorling Kindersley

Produced for DK by
COBALT ID
www.cobaltid.co.uk

Managing Editor Marek Walisiewicz
Managing Art Editor Paul Reid
Art Editor Darren Bland

DK LONDON
Project Editor Amy Slack
Managing Editor Ruth O'Rourke
Managing Art Editors Christine Keilty, Marianne Markham
Production Editor Heather Blagden
Production Controller Stephanie McConnell
Jacket Designers Nicola Powling, Amy Cox
Jacket Co-ordinator Lucy Philpott
Art Director Maxine Pedliham
Publisher Katie Cowan

First published in Great Britain in 2022 by
Dorling Kindersley Limited
DK, One Embassy Gardens, 8 Viaduct Gardens,
London, SW11 7BW

The authorised representative in the EEA is
Dorling Kindersley Verlag GmbH.
Arnulfstr. 124, 80636 Munich, Germany

Copyright © 2022 Dorling Kindersley Limited
A Penguin Random House Company
10 9 8 7 6 5 4 3 2 1
001–326193–Jan/2022

A CIP catalogue record for this book
is available from the British Library.
ISBN: 978-0-2415-3053-5

Printed and bound in China

For the curious
www.dk.com